THE EAGLE'S NEST

History of American Science and Technology Series

General Editor, LESTER D. STEPHENS

The Eagle's Nest: Natural History and American Ideas, 1812–1842, by Charlotte M. Porter

THE EAGLE'S NEST

Natural History and American Ideas, 1812–1842

CHARLOTTE M. PORTER

The University of Alabama Press

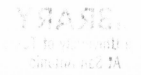

Copyright © 1986 by
The University of Alabama Press
University, Alabama 35486
All rights reserved
Manufactured in the United States of America

Publication of this book was made possible, in part, by financial assistance from the
Andrew W. Mellon Foundation and the American Council of Learned Societies.

Library of Congress Cataloging-in-Publication
Data

Porter,
    Charlotte M., 1948–
    The eagle's nest.

(History of American science and technology
series)
    Bibliography: p.
    Includes index.
    1. Natural history—United States—History.
I. Title. II. Series.
QH104.P67 1986   508.73'09   85-16465
ISBN 0-8173-0280-8

*To my parents, whose ideas continue
to surprise their children*

Contents

vii

PART FOUR: PUBLICATIONS

Illustrations

Acknowledgments

The voluminous archival resources and manuscript holdings scattered across the United States challenge the most zealous student of American science and ideas. Fortunately, my research was made more rewarding by the assistance of gifted librarians, Stephen Catlett of the American Philosophical Society, Sylva Lewis and her staff at the Philadelphia Academy of Natural Sciences, and Nina Root, who, over the years, has relieved the long hours I spent in the Library and Rare Book Room of the American Museum of Natural History with well-placed suggestions and a reliable sense of humor.

In addition, my requests were given careful attention at libraries of the Missouri Botanical Garden, the New York Botanical Garden, Fort Leavenworth, and Platte County, Missouri. Librarians at the Spencer Library at the University of Kansas, the Linda Hall Library in Kansas City, the New York Public Library, and the New-York Historical Society were especially well-informed about their respective holdings and shared their hunches, hopes, and experience with me. I also appreciated access to research materials at the Metropolitan Museum of Art and to archives of the Smithsonian Institution and the American Museum of Natural History.

At Harvard University, Everett Mendelsohn, professor of the history of science, guided early stages of my research and writing.

Ann Blum, archivist at the Museum of Comparative Zoology, was a bright guide to the formidable natural history holdings at the museum and other parts of the university. Research at the Houghton Library and the Library of the Gray Herbarium was always fruitful, but perhaps my happiest moments were those spent with Raymond Paynter, who kindly showed me the Peale museum species housed in Harvard's Department of Ornithology. "They are easy to identify," he said, and indeed they were. Other specimens lay quietly in their drawers like stuffed feathered socks, but Peale's birds looked alive and ready to rejoin the late eighteenth-century landscape of their creator's fertile imagination.

I am also grateful to Ralph Schwartz, director of New Harmony, Inc., and to Charles Boewe, editor of the Papers of C. S. Rafinesque, for their helpful personal communications, which are acknowledged in the footnotes. Conversations with Edgar Richardson, Whitfield Bell, Edward C. Carter II, Nathan Reingold, Marc Rothenberg, Samuel Proctor, and Ernst Mayr have provided more insights than they may realize into the history of American ideas and natural history. Equally important, Fred G. Thompson, curator of malacology at the Florida State Museum, introduced me to the ecology of southeastern species described by the early nineteenth-century naturalists I discuss. Last but not least, a horse named Valiant enabled me to follow parts of William Bartram's route in Florida and to see some of the same sights that Quaker naturalist saw so many years ago.

I owe special thanks to the Director's Office of the Florida State Museum for funds that facilitated the preparation of this book. The steadfast clerical support of Jennifer Hamilton and Cheryl Wilson far exceeded my expectations, and without critical readings of the text by both Sally Kohlstedt and Lester D. Stephens, this work would not have achieved its final form.

THE EAGLE'S NEST

Introduction: An Invitation

In 1853, Baldwin Möllhausen, a self-taught German artist traveling with the Pacific Railroad Survey under Lieutenant Amiel W. Whipple, complained that "the want of sense of the beautiful in nature in the generality of Americans, is to a European as remarkable as the enthusiasm of the European is to them absurd."[1] If Möllhausen's remark was true for Americans of the mid-nineteenth century, it did not hold for an earlier generation of naturalists who explored the West. They approached the observation of nature with an enthusiasm quickened by the emotions that had characterized the War of 1812, and despite the primitive conditions for scientific study imposed by frontier travel, they developed a sense of the beautiful in nature which was as non-European as it was scientific.

In June of 1804, Alexander Wilson (1766–1813) published "The Invitation" in the *Literary Magazine and American Registrar*. This long verse suggested that the reader leave the bustle of Philadelphia to join Wilson on

> One kind excursion. . . .
> Come, then, O come! your burning streets forego,
> Your lanes and wharves, where winds infectious blow,
>   Where sweeps and oystermen eternal growl,
> Carts, crowds and coaches harrow up the soul.

Each summer, prosperous residents abandoned the city to the ravages of yellow fever, but Wilson had more in mind than a retreat to the beautiful suburbs of Germantown or Queen Lane. He promised his companion "deep, majestic woods, and op'ning glades, / and shining pools, and awe-inspiring shades." Four years later, he made good his promise and presented his reading public with the "old associates of his lonely hours"—the illustrated birds of his *American Ornithology,* a book that heralded a new era for natural history in the United States.

As he moved toward the completion of this nine-volume work, Wilson continued to write popular nature poetry. Other naturalists traveling on the frontier also responded to the particular qualities of the North American landscape in which they worked. In their journals, they recorded the notable features of the terrain as well as their reactions to those vistas. As a result, passages of landscape description took up a significant portion of their journals. Many elements of their descriptions were not new to readers familiar with the British Romantic literature of the early nineteenth century.[2] Like the more gifted Romantic poets William Wordsworth and Samuel Coleridge, the American naturalists described the irregularities of the untamed landscape—giant shadows cast by moving clouds, rocky turrets, roaring cataracts, and other "stern features of nature." American readers responded because naturalists out west could physically experience the elements Edmund Burke associated with the complex aesthetics of the sublime: difficulty of access, a feeling of privation, magnificence, vastness, and even the infinite.[3] Thomas Say (1787–1834), Thomas Nuttall (1786–1859), Titian R. Peale (1799–1885), and John K. Townsend (1809–51) devoted the best part of their careers to the study of western natural history. John James Audubon (1785–1851) shared a similar interest; early in his career he outlined his "ardent Wish" to travel west to explore the Arkansas Territory.[4]

These naturalists' understanding of the frontier landscape differed from that of the Romantics in at least one important way. Instead of regarding the wilderness as "incomprehensible" or "awesome," they discerned in the "deep romantic chasm" a geological system. They gathered taxonomic specimens in the "wild secluded" scene, and behind the "purple shadow" they discovered new genera and species.[5] Like their counterparts abroad, these early nineteenth-century travelers applied the tools of classification to new flora and fauna with remarkable success. As a consequence, during the three decades that followed the War of 1812 in the United States, systematic study of natural history emerged as an organized pursuit. The transition from the independent efforts of individuals to institutional control of collections and publications was not easy, and the reputations of key figures did not survive the changes they had set in motion. In Philadelphia, the nation's leading city of science, one group founded the Academy of Natural Sciences, an organization that eventually counted among its members Gerard Troost (1776–1850), C. S. Rafinesque (1783–1840), and Charles Lucien Bonaparte (1803–57), as well as Say, Nuttall, Peale, and Townsend. These men attempted to emulate Wilson, for his *American Ornithology* immediately set the standard against which all other works of natural history would be judged.

Soon after the war ended in 1815, the arrival of such gifted foreign-born naturalists as Nuttall, Rafinesque, Charles Alexandre LeSueur (1778–1846), and Bonaparte combined with the remarkable patronage of William Maclure (1765–1840) to create a group enthusiasm for American natural history. Research at the newly incorporated Academy of Natural Sciences rapidly expanded to include entomology, conchology, mammalogy, botany, and even ethnology in addition to the initial and persistent interest in birds. Even though both of the academy's presidents from 1812 to 1840, Troost and Maclure, were mineralogists, the dominant concerns of the membership continued to be zoological. As a consequence, Nuttall and Rafinesque, who also worked extensively with plants, tended to approach botany with ideas derived from animal studies.

Although efforts to establish an American approach to natural history met with initial acclaim, there were two technical problems. First, like many of their contemporaries, academy authors found the Linnaean system of classification for plants and animals disappointingly artificial. The Linnaean method popularized natural history in the classroom because it was easy to use, but academy writers believed the Linnaean system limited their scientific contributions. They anticipated the adoption of a more "natural" method of classification, but they also recognized that such a system would disrupt taxonomic continuity in the literature and necessitate wide-scale educational reforms

The basis for taxonomic natural history is type specimens or the individuals upon which the taxonomic designation is based. That many type specimens for North American species were housed abroad presented a second problem for naturalists working in the United States. Lacking access to these valuable foreign reference materials, authors relied upon their own collections. As a result, their critics accused them of creating self-serving taxonomies. With the trans-Mississippi West wide open for their collection needs, however, emigrés such as Nuttall and Troost or political refugees such as Wilson or Rafinesque were not about to return to Europe to perform museum research on old and often crudely preserved materials. This situation led to the troublesome synonymity and confusion in their early published papers, which divided the natural history community by the 1830s. Critics demanded a moratorium on publications, and in 1837, a young doctor from New York State named Asa Gray (1810–88) wisely planned a visit to European collections that secured his position of leadership within the American scientific community.

A third factor that determined the success of early academy members in Philadelphia was financial. Maclure furnished the academy with everything from scientific reference works, laboratory equipment, and printing facilities to the buildings in which it was located, but his patronage was a two-edged sword. By 1826, the institution's membership was divided over New Harmony, a utopian village in Indiana purchased by the British socialist Robert Owen and financed by Maclure. The first Harmonie community had been built on the banks of the Wabash River by a strict but versatile group of German Pietists. Owen, by contrast, was unabashedly atheistic as he sought an American counterpart to New Lanark, his model city of mill workers in Scotland. Maclure hoped to implement a great social experiment by educating orphans free from parental prejudice. Natural history instruction without religious bias was in his view essential to this process. The motives of those who accompanied Maclure to New Harmony are less obvious. Say and LeSueur were full-time field researchers, who relied on Maclure's munificence to subsidize their travels and often their living expenses.[6] When Maclure moved to New Harmony the other teachers who accompanied him were also in his employ. Troost soon quit the community, but Say, LeSueur, and John Speakman remained. Despite Maclure's advancing years, his ideas continued to influence those around him. One of the best-educated mineralogists in the country, Troost did not return to Philadelphia after he left New Harmony. At the frontier University of Tennessee he established an important center in the Southeast for geological education and surveys. His accomplishments paralleled the work of Owen's sons, Maclure's intellectual heirs, at New Harmony.

Maclure's new "science of society" was actually an idiosyncratic economic geology, and although his personal interest in natural history was not taxonomic, he promoted systematics through his patronage of others. In Philadelphia, Say and his colleagues had discussed the need for a more natural system of classification, but they did not implement such a system. As superintendent of education at New Harmony, Say was in a position to make reformed natural history a reality for the education of a new kind of natural history reader was possible at Maclure's schools. Introduction of a "true" method for classification became a practical goal through Maclure's so-called Education Society and not merely an intellectual pastime of conversant academicians.

Other members of Maclure's generation also influenced the Philadelphia academy. By 1791, Say's great-uncle William Bartram (1739–1823) had articulated attitudes toward nature and living things that younger naturalists eagerly assimilated. Bartram's descriptions of southeastern landscapes he traversed from 1773 to 1777 created a uniquely American literary masterpiece. Bartram was also a visionary, and his book of travels describes nature as a beautiful, vulnerable state in which democracy finds precedent in the relationships of plants and animals. Bartram embarked upon his twenty-four-hundred-mile trip as a British subject. He returned to the Philadelphia of his youth as the citizen of an emerg-

William Bartram (1739–1823), 1808. Charles Willson Peale painted the Quaker botanist long after his famous trip of 1773–77, but the artist has referred to his colleague's achievement with a sprig of *Jasminium officiale*, a flowering vine of the Southeast. (Independence National Historical Park Collection)

ing nation. With time, his understanding of nature was reshaped by nationalistic ideals. Like many visitors to Florida, he had enjoyed watching the fish in the pellucid spring waters near Lake George. "Here," he wrote, the "element in which they live and move is so perfectly clear and transparent, it places them all on an equality with regard to their ability to injure or escape one another." He completed his image of American social mobility: "The trout freely passes by the very nose of the alligator and laughs in his face, and the bream by the trout."[7]

In 1817 in one of his first actions as president of the academy Maclure organized a scientific party to retrace Bartram's "track" in Florida. The opening pages of Bartram's book, by then a well-read classic, urged new attitudes about the intellectual abilities of American aboriginals. For Bartram, the ability to adapt which he had observed among animals provided both a zoological basis for human education and his ultimate argument against the French school of thought that viewed beasts as living machines.[8]

Bartram's influence was not confined to his writing. Wilson regarded him as his first instructor; Say and young Titian Peale were frequent visitors, and Nuttall was given a room at the Bartram family's famous gardens in Kingsessing. Indeed, the second part of the introduction to a bird manual Nuttall published a decade after Bartram's death is really an essay on Bartram's thesis. Nuttall claimed that birds, capable of "education, or the power of adding to their stock of invariable habits," were not "animated machines," but possessed "an incipient knowledge of cause and effect." Nuttall discussed the "frequent demand" in nature for "relieving invention."[9] Although these conclusions were in diametric opposition to the thought of William Buckland fashionable in some British circles, the analogy between animal modifications and human education as a means of self-improvement remained attractive for many Americans.

The patriotic focus of naturalists working at the academy in Philadelphia is evident in their fundamental tenet that the species of North America were unique and distinct from European counterparts. In the sixth query of his slim *Notes on the State of Virginia,* Thomas Jefferson repudiated Georges Louis Leclerc (1707–88), comte de Buffon's theory of degeneration in the New World. In his defense of American nature, Jefferson suggested that the species were not degenerate or derivative forms of Old World animals but on study would prove to be distinct species. Like his friend Bartram, he continued to encourage younger men to verify this thesis, but his enthusiastic support of science and exploration also made him the target of political satire even after the unqualified success of the Lewis and Clark Expedition.

William Cullen Bryant (1794–1878) vilified the embargo in his first published poem with all the antiscientific sentiment a precocious thirteen-year-old boy could muster:

> Go, Wretch, resign the presidential chair,
> Disclose thy secrete measures, foul or fair,
> Go, search with curious eyes for horned frogs,
> 'Mongst the wild wastes of Louisianian bogs;
> Or where the Ohio rolls his turbid stream
> Dig for huge bones, thy glory and thy theme.[10]

The youth had perceptively identified Jefferson's "theme." As an older Hugo Meier has noted, the halls and walls of Monticello, its "architectural design, its kitchens, workshops, and household furnishings, all reflected the diversity" of Jefferson's interests.[11] The horned "frog" was eventually sent to Charles Willson Peale (1741–1827) to draw, and Jefferson's promotion of American collections reached its zenith in the Philadelphia Museum, which Peale opened to the public in 1786.[12] Twenty years later, his museum housed the first reconstructed mastodon, mounted mammals from the Lewis and Clark Expedition, and a menagerie of living animals forwarded to his care by merchants, travelers, and naturalists from all over the world. At least one viewer exclaimed that a visit to the museum was worth the trip across the Atlantic. Anne Royall was most impressed by a stuffed sea lion: "Even the eye lash was entire."[13] Besides sightseers, young artists and naturalists (including Peale's talented sons and daughters named after famous painters) frequented the museum's rooms to draw and study exotic animals, human skeletons, and costume displays.

Peale's museum and the publications and paintings it inspired reflected an age of optimism, and throughout his long life Peale retained a cheerful confidence in the goodness of nature. In 1809, his *Epistle to a Friend on the Means of Preserving Health, Promoting Happiness and Prolonging the Life of Man to Its Natural Period* was printed for distribution, and American ideas about human health and the welfare of a new nation, remained bound up with an interpretation of nature. According to Jefferson, Buffon had argued that man-made modifications of the environment could halt or at least allay the process of biological degeneration in the overly moist climate Buffon attributed to the New World.[14] Since other proponents of the theory of degeneration emphasized that the indigenous peoples of North America did not possess sufficiently advanced cultures to carry out these desirable changes, the real question was the role of European inhabitants as caretakers of nature and her "denizens," the Indians. Yet, as surely Jefferson and his most thoughtful readers realized, at the same time that the new white Americans undertook agricultural improvements they introduced a most debased means to implement these changes—slavery. By 1819, the classification of the human races in a way that was consistent with the rest of American taxonomy became the most difficult challenge acknowledged by naturalists who believed in the uniqueness of North American species.

During the period 1812 to 1842, the problems of human classification, speciation, type specimens, and financial support widened the inevitable separation of botanists and zoologists into two well-defined groups. Colonial botanists such as Alexander Garden, William Bartram, or his father, John Bartram, had worked closely with planters and pharmaceutical businesses. By contrast, zoologists working in early nineteenth-century America did not identify with the resident community. This divergence of botany and zoology is illustrated by the descendants of John Bartram. Within four generations, this productive family of naturalists had moved from royal to federal connection and from botanical avocation to zoological specialization. Eventually botanist to George III, self-taught John Bartram (1699–1777) corresponded and exchanged specimens with Peter Collinson and other British Quakers. Like his cousin Humphrey Marshall, he was a successful nurseryman, and he established commercial gardens at Kingsessing outside Philadelphia. Upon his death, they passed to his oldest son, John, and then to John's daughter, Anne Carr. William Bartram became a pivotal figure in the development of both botany and zoology. William declined Jefferson's request to become the first naturalist assigned to a United States government expedition in 1803 and recommended Alexander Wilson instead.[15] Wilson was not chosen, and the first naturalist officially appointed to a military expedition was Thomas Say, the great-nephew of William Bartram's twin sister. Say, who lost a considerable inheritance through an ill-advised business foray with John Speakman, initiated his own scientific career without foreign sponsorship. An entomologist, he became an early pillar of the Academy of Natural Sciences, of which his wife, Lucy Way Sistaire Say, became the first female member in 1841.

Jefferson believed that all forms of science would prosper under a republican government. Maclure's utopianism rested upon the same belief, and William Bartram repeatedly described areas in East Florida as "paradise." Paradise of the biblical account, of course, was not a wilderness but a garden where innocent man had lived at peace with the beasts he named. The plants flourished, and the animals wandered harmlessly among the vegetation. Later naturalists from the academy viewed nature in much the same way. Their view of nature was a quieter or, to use Wilson's term, "humane" mode of thinking pervaded by a deep sense of nationalism. As an eighteen-year-old boy assisting Say on the Missouri River, Titian Peale expressed the appealing elements of their scientific sentiments with his nicely phrased observation of "a goose in peaceable possession of an eagle's nest."[16] As an older man en route to Rio de Janiero with the first United States Exploring Expedition of 1838, he described a day at sea: "We are now rolling and plunging merily (*sic*) before a strong trade wind, with sudden sails below and aloft, on both sides, in a way that surpasses any Eagle with spreading wings, designed for coin."[17] Peale, the foremost American painter of birds next to Audubon, no doubt envisioned science like this ship, which carried half of the expedition's natural history personnel, and this invigorating image may have inspired the dies for the eagle coins he designed for the United States Mint in 1848.

Peale's patriotic associations were not unique. Amos Eaton (1776–1842) wrote to John Torrey (1796–1873) in 1822, "You are made for the highest walks of science . . . to correct the blunders of others and to keep the ship of science in trim."[18] Peale's ship of science, the *Peacock,* however, was a badly leaking sloop-of-war that was wrecked in the mouth of the Columbia River. By the time the expedition returned in 1842, Peale's attitudes were no longer considered good science in professional circles.

In the meantime, Torrey and his student at the New-York Lyceum of Natural History, Asa Gray, endeavored to "keep the ship of science in trim." They introduced new methods for botanical classification, and, equally important, they introduced professional standards that eventually excluded early members of the Academy of Natural Sciences in Philadelphia, who were explorers. Maclure's ideas prospered at New Harmony, by midcentury an important center for government-sponsored geological surveys, but for the most part, American natural history became an increasingly conservative intellectual pursuit until the advent of Darwinism.

PART ONE

Precedents

The Quality of Nature

The two giants of eighteenth-century natural history were Carolus Linnaeus (1707–78) and Georges Louis Leclerc, comte de Buffon. Born in the same year, each of these men developed a very different approach to the study of nature. Linnaeus sought out the diversity of living things and perceived the organization of this diversity through science as a human obligation. Buffon's view, larger and less practical, encompassed eons of geological time and a universal outlook in which numbers, weight, and measure were the means to knowledge. Buffon saw living things as the results of an extremely ancient historical process. Linnaeus, by contrast, considered them to be physical units of a divine plan determined by their bodily parts. Although neither man crossed the Atlantic, both used the collections and correspondence of others to write about American species. Linnaeus revolutionized the basic rules naturalists used to describe plants and animals. Buffon's contributions were more philosophical and more controversial.

A brilliant stylist, Buffon began publishing his massive *Histoire naturelle, générale et particulière* in the middle of the eighteenth century or slightly after the publication of Linnaeus's *Systema naturae* in 1735. By the time of his death in 1788, Buffon was one of the most widely read authors in France. His writings went through six editions, and his ideas were widely known in the British colonies.

15

The Philosophical Society founded by Benjamin Franklin in Philadelphia invested in his costly volumes, and by 1800, several less expensive edited translations were available for English-speaking readers.[1] Buffon denied the validity of the basic Linnaean classification of living things by genus and species. In 1766, he argued that many of the species designated by the Linnaean system could be reduced to a smaller number of original types from which they all had developed. For quadrupeds that number was thirty-eight. Buffon's basic criterion for a species was the ability of interbreeding organisms to produce fertile offspring. By midcentury, this definition was included in French dictionaries, and it is still used in modern school texts.

Arguing that the earth had undergone a long history of change, Buffon postulated that forms could be spontaneously generated from organic molecules in the environment without a special act of creation or divine mediation. Chance and chemical affinities brought organic particles together. Because only the most powerful and most complete molds persisted, the new creatures that resulted, Buffon argued, resembled the living forms now familiar to zoologists. Because their generation had occurred at a lower temperature in a cooling world, the more recent creatures differed in one important way from the original forms: the new animals were diminished in size or degenerate. Buffon did not stop here. The recorded similarities of animals in the northern regions of both the Old and New Worlds led him to postulate the earlier existence of a land bridge between the continents of Asia and America, and he suggested that the fauna of the New World, which he viewed as smaller, less active forms of Old World species, had spontaneously generated from scattered *moules*. This process of degeneration, he claimed, continued in the New World because of the moist, unhealthy climate. Unlike his predecessors, Buffon related the concept of species to geography and the history of the earth, which he dared to argue was of far greater antiquity than commonly acknowledged. With his theory of the seven epochs of earth, Buffon introduced the concept of time into natural history and made possible an entirely new set of relationships for nineteenth-century naturalists to investigate.[2]

Although the colonial naturalists such as Alexander Garden condemned Buffon's criticism of Linnaean divisions as arbitrary and found the materialism of his organic molecules repugnant, they were reluctant to enter into disputes with foreign correspondents on whom they depended for scientific news.[3] Their silence came to an end, however, in 1776, when a nineteen-page article on America prepared for the new supplement to the French *Encyclopédie* applied Buffon's arguments to both the North American Indians and the colonial settlers. The author was a Hollander named Cornelius de Pauw, who with typical eighteenth-century *ésprit de système* elaborated Buffon's hypothesis for man. Acting in an era in which underpopulation was perceived as a greater threat to human happiness than overpopulation, the French government disseminated de Pauw's notions as propaganda to discourage emigration to the new United States.[4] De Pauw's article in turn encouraged American wits to sharpen their pens. For example, Benjamin Smith Barton (1766–1815) quipped that according to de Pauw's logic the reason native North American bees are solitary rather than social was "weakness of political union" in the New World.[5]

After 1778 eight volumes of Buffon's *Histoire naturelle des oiseaux* were published, bringing the total to nine volumes of brilliant prose with spectacular illustrations of birds. By 1783, the challenge posed to the newly recognized United States was no longer confined to humorous remarks, and the first and perhaps the most comprehensive American effort to reexamine the theory of degeneration was Thomas Jefferson's *Notes on the State of Virginia*. Written in 1781 and circulated in Paris in French, this popular little book went through at least fourteen printings and four editions from 1788 to 1832. Jefferson addressed Buffon's claim that the quadrupeds of the New World were physically deteriorated versions of Old World forms.[6] Repeating the comparisons of body weight given in Buffon's text, Jefferson demonstrated that various American mammals were as large as, if not larger than, comparable European mammals. Ignoring Buffon's treatment of biological diversification, Jefferson showed that Buffon's data did not support his theory.[7] Jefferson turned to paleontology for the grandest example that the natural forces of the New World were still vigorous.

He claimed that the North American mammoth (really a mastodon) "should have sufficed to have rescued the earth it inhabited and the atmosphere it breathed, from the imputation of impotence."[8]

Ironically, the mammoth raised the thorny issue of extinction, a real problem for Jefferson's thesis of natural vigor. The disappearance of so large a species supported rather than contradicted Buffon's picture of an enfeebled North American nature, and throughout his life Jefferson would not admit the possibility of natural extinction.[9] In the *Notes* he argued rather weakly that although the mammoth had been exterminated by Indian hunters in eastern North America, it might still be found in the undisturbed continental interior.[10] Jefferson's scientific colleagues endorsed his anti-Buffonian sentiments, but they denied Jefferson's awkward explanations of fossil forms. In an obituary of Jefferson read before the New-York Lyceum of Natural History in 1825, Samuel Mitchill (1764–1831) praised Jefferson's promotion of American science with an important aside on his own belief in extinction, which concluded: "The individuals of a race drop off until none survive."[11]

Jefferson's success in the *Notes* rested upon Buffon's poor choice of zoological examples.[12] Although Jefferson defended animal anatomy of the New World, he ignored the significance of faunal similarities. By his reliance on physical parameters, namely body size, Jefferson reflected the eighteenth-century love for numbers, weight, and measure, and he must have taken special pleasure in Sir Isaac Newton's mention of nature in America in the "Rules of Reasoning in Philosophy" in Book III of the *Principia*. The *Notes* motivated that group of eighteenth-century professionals who were the most attentive to the quality of the American environment—physicians. Reflecting the ancient Hippocratic concern of *Airs, Waters, Places*, the nation's leading doctor, Benjamin Rush, published his "Account of the climate of Pennsylvania and its influence upon the human body" in 1790. Two years later, in a four-hundred-page work, William Currie, a fellow of the College of Physicians of Philadelphia, attempted to relate climatic factors, health, and recent colonization. Rush and Currie distinguished "clearing" from "cultivating," and both remarked that in Pennsylvania increased sickliness was not a natural consequence but followed settlement, deforestation, and the creation of millponds.[13] Soon naturalists joined physicians in their defense of the North American landscape as a suitable habitat for man and beast.

William Bartram's familiarity with Buffon's thesis is apparent in his manuscript Commonplace Book and throughout his long, rambling account of his travels in the Southeast, which was published in 1791.[14] Bartram frequently paused to remark upon "sublime enchanting scenes of primitive nature" or "visions of terrestrial happiness," although his early efforts to run an indigo plantation outside St. Augustine in 1767 had failed miserably. Bartram made a point of noting that the Indians managed their environment. The "trees and shrubs which cover these extensive wilds were not stunted by any barrenness of the soil," but, he argued, were "kept down by the annual firing of the desarts." Furthermore, in Florida's Alachua Savanna (present-day Payne's Prairie State Preserve), Bartram described the cattle of the Lower Creeks as "large and fat as those of the rich grazing pastures of Moyomensing in Pennsylvania."[15] After admiring the "vast size and strength" of the "animal creation" and questioning the "General opinion of philosophers," which has "Distinguished the moral system of the brute creature from that of mankind," Bartram went on to consider the "savage" character. Was it animal or human? Were the Indians "deserving of the severe censure, which prevailed against them among the white people, that they were incapable of civilization"? Could they "adopt the European modes of civil society"?[16] Bartram was satisfied that the Indians were "desirous of becoming united with us, in civil and religious society." He suggested that the federal government send envoys "as friendly visitors" to implement this transition as soon as possible, for "our negligence, in the care of the present and future well-being of our Indian brethren, induces me to mention this matter."[17] Bartram's writings identified nature with social issues, and other naturalists followed parts of his route to rediscover his sensibilities through firsthand experience with his sources: Alexander Wilson in 1809, Thomas Nuttall in 1815, William Baldwin in 1817, William Maclure, Thomas Say, George Ord, and Titian R. Peale in 1817–18, John E. LeConte in 1822, and perhaps Bartram's only harsh critic, John J. Audubon, in 1831.

As with Bartram, the relative roles of the environment versus human action in the health and "well-being" of living creatures in the New World remained a recurrent theme in American circles of science, and writers continued to defend the progress of human populations in the United States. In a long published letter to the brilliant Philadelphia clockmaker, David Rittenhouse, William Barton attempted to correlate quantitatively climate, manners, and health with Benjamin Franklin's famous remark that in America the population doubles every twenty-five years.[18] Barton's arguments repudiated de Pauw's thesis at every turn, and his interests shifted from human issues to natural history, species, and species numbers. Barton's botanical postscript literally contained the seeds for a new approach to American natural history, and by 1818, naturalists were publishing systematic works in which American species were recognized as distinct from Old World forms.

In the works of natural history published in Philadelphia after Bartram's *Travels,* the phrase "vain philosophy" became virtually synonymous with Buffon's theory of degeneration. Even the word "philosophical" was used with utmost discretion, a reaction reinforced by an older attitude toward ornamental or unserviceable knowledge.[19] Alexis de Tocqueville's observation that those in America "who study the sciences are always afraid of getting lost in utopias" at first sight seemed to be true as American natural history gained practical expression.[20] Even in an age when the beauty of machines was admired beyond any measure of utility, eighteenth-century French *philosophes* had criticized biology based upon the Cartesian dualism of the human mind and animal body.[21] Like Bartram, a younger generation of naturalists in the United States rejected the position that animals were mindless biological machines, but they did not object to all of Buffon's ideas. His "northern principle," which claimed a common biology for Canada and the British Isles on the basis of a former land bridge, was generally accepted, and Buffon's work remained popular among British readers throughout the nineteenth century.[22]

Although no one used the term "degeneration," American horticulturists had long recognized a limited plasticity in the makeup of botanical species. Many, including Jefferson and Bartram, considered the garden so cherished by eighteenth-century amateurs as a slowly paced experiment in hybridization, and these Americans were not distressed by the idea that plants deteriorated or improved in different natural environments. In fact, Philadelphia botanists made it their business to learn just how they could affect that process in gardens maintained by the Bartram family, Humphrey Marshall, David Hosack, and Bernard M'Mahon. As a consequence, horticulture, that loveliest expression of American economic botany, was never Americanized for its very existence depended upon a transatlantic exchange.[23] Horticulturists did not consider a nation's species among its inalienable possessions, and they sent plants, seeds, and stocks to be shared, modified, and made famous in other lands.

By the early nineteenth century, this attitude about American species distinguished horticulturists from naturalists with taxonomic interests. Horticulturists wanted to export seeds of new, exotic American species; taxonomists did not wish to be preempted in description by foreign authorities. Furthermore, many of Buffon's arguments made sense when applied to plants rather than animals. In the *Notes,* Jefferson had interpreted degeneration to mean a deterioration of organic bulk or biomass. Just as land was evaluated on the basis of bushels harvested, Jefferson suggested, Buffon had evaluated the quality of nature in North America by the animal weight a given locality sustained. Had Jefferson's agricultural interests affected his exposition of Buffon's ideas? Although their conclusions differed, the parallels between environmental management of species and husbandry were evident in both Buffon's and Jefferson's positions, and the validity of those parallels remained unquestioned either by early readers of the *Notes,* those Virginia landholders who were Jefferson's close friends, or by later readers with no agricultural expertise.[24]

Buffon may have derived some of his ideas from agriculture, for as a young man he had pursued experimental silviculture on his estate in Burgundy. Buffon's theory of degeneration, which hinged on questions of human management and climate, did lend itself to forestry, and not illogically, Buffon's countryman François André Michaux (1770–1855) attempted to evaluate North American trees by using Buffon's methods for mammals. First published in Paris from 1811 to 1813, Michaux's beautifully written work, illustrated by the "Raphael of the rose," Pierre-Joseph Redouté (1759–1800), and his brother, was promptly translated under the English title, *North American Sylva*.[25]

Unlike Buffon, Michaux was familiar with his subject, and he had access to his father's valuable botanical notes from 1785 to 1796. Having been a political observer in Persia, André Michaux had continued to serve the French government quietly in this capacity during his North American excursions. He and his son first came to the United States in 1785 to select trees for transplantation to French soil and, in 1787, they purchased a plantation outside Charleston, South Carolina. The older Michaux traveled to the Bahamas and to Canada, and in 1794, on Jefferson's urging, the American Philosophical Society undertook a subscription for Michaux to lead a botanical trek throughout the Southeast. Jefferson put up £50 for the venture, but the trip was quietly aborted after the members of this august group learned that the French minister had engaged Michaux to stir up Kentuckians against Spanish control west of the Allegheny Mountains. Michaux may have known more about plants than any other man in America, but he was also a spy.

Meantime, François André Michaux had returned to France for medical studies. After his father's death in 1802 and a second trip to the United States, Michaux prepared notes salvaged from the shipwreck of the vessel on which his father had been sent back to Europe in 1796.[26] The elegant book that resulted from their joint efforts was critical of American attitudes about the management of their biological resources. Expressing the mechanistic taste of the late eighteenth-century Enlightenment, Michaux *fils* claimed his interest in American forest trees was the "utility of each species in the mechanical arts."[27] Michaux's evaluation of American timber was based on a quality he called elasticity, the weight required to bend the wood. He admitted that this criterion might be misleading because American woods were not properly seasoned before use and hence did not test well, but even this concession to the trees placed the American practices in a bad light.[28]

Like Buffon, Michaux associated the "extreme unwholesomeness of the climate" with "tertian fever" or disease. He claimed that by early fall everyone looked so sickly that "Georgia and Lower Carolina resemble, in some measure, an extensive hospital." The trees he described growing under such conditions formed "squalid woods." The upland willow oak was "one of those abject trees" found on lands "abandonned in account of their sterility." The wild bear's or Banister's oak was "an infallible index of barren soil." Similarly, the diminutive, hence useless, chestnut oak was "a certain proof of the barrenness of the soil." Michaux attributed the relative scarcity of the valuable American white oak to the "dryness or humidity" of the soil and recommended the introduction of the commercially superior European tree. An early advocate of conservation, Michaux urged that the federal government support "all means of preserving and multiplying" exhausted species. He even suggested that endangered species be imported to France, where they might have some chance of surviving under proper management.[29]

Michaux had maintained a cordial correspondence with William Bartram and had even written Alexander Wilson to ask whether his publisher, Samuel Bradford, would accept his book, and under the auspices of the second president of the Philadelphia Academy of Natural Sciences, William Maclure, the French edition was distributed among various agricultural societies in this country.[30] An edition was reprinted from the original copperplates at New Harmony, Indiana, and the Academy of Natural Sciences in Philadelphia published a popular supplement by Thomas Nuttall, which went through at least five printings. After the Civil War, the academy issued another deluxe edition. A crusader, Michaux deeded his father's American manuscripts to the American Philosophical Society with a special fund that led to the establishment of Arbor Day. Yet these actions do not explain the immediate and lasting American endorsement of Michaux's *Sylva*.

The theory of degeneration seemed to encourage American horticulture, and silviculture actually benefited from Michaux's critical attention and gained new inroads with general readers. At the New-York Lyceum of Natural History, an organization too poor to entertain the idea of a fine library or major collections, botanists turned once again to Europe for intellectual leadership, ongoing correspondence, and new ideas. For them, the "Book of Nature" was not written in a distinctly American language, and they valued foreign authorities more than the opinions of their American colleagues. This international shift was certainly a necessary development within American natural history and one Jefferson would have applauded, but ironically it met with the greatest resistance from those who justified their own ideas with Jefferson's ultimate refutation of Buffon, the discovery of new species in the New World. In 1815, DeWitt Clinton warned his constituents in New York State that in the area of science, foreign colleagues "view us with a sneer of supercilious contempt." The reason Clinton gave for this sorry treatment was Buffon's declaration that "in America, animated nature is weaker, less active, and more circumscribed in the variety of productions, than in the old world." Clinton continued somewhat effusively that foreign scientists claimed "our national character is marked with all the traits of premature corruption and precocious turpitude."[31]

Although these comments must have sounded anachronistic to a younger generation of naturalists, Clinton was attempting to give new life to Jeffersonian sentiments that had promoted science. Seven years later, in a newspaper column published under the pseudonym "Hibernicus," he set out to provide a New York equivalent to Jefferson's *Notes on the State of Virginia*. Like Jefferson, Clinton complained that American species of birds were "lost under European names." He advocated statewide research of plants, animals, and mineral resources, studies practical for the first time along the new canal excavated across the state during his gubernatorial administration, and he praised the New York City fish market as "pregnant with instruction and amusement." In no uncertain terms, he reproached antiscientific "graciosos or buffoons" and assured his readers that "the most virulent accusation against Jefferson is the impalement of butterflies."[32] Alexander Wilson had traveled throughout New York, and after 1804 his

serial poem "The Foresters" made the state's main natural attraction, the falls of Niagara, as well known in the parlors of Philadelphia as Jefferson's description of Virginia's Natural Bridge.[33]
Clinton was well aware of Wilson's work and its roots in Jeffersonian thought, and he hoped his column would encourage others to
follow Wilson's example. Clinton foresaw the time when natural
history would not be confined, in Charles Willson Peale's words,
to the "industry, patience and zeal of an individual," and he shared
the artist's optimism in "the tendency which a knowledge of
Natural History has to promote National and Individual Happiness."[34]

Although his efforts immediately attracted only C. S. Rafinesque, who eagerly dedicated his hastily written *Florula Ludovi-
ciana* to Clinton, the governor clearly anticipated the founding of a
New York State museum, for the remains of the first reconstructed mastodon in Peale's Philadelphia Museum had come from Orange County, New York. Like Maclure, Clinton also perceived the
future relationship between natural history collections, geological
surveys, and commerce. Clinton's advocacy of natural history was
somewhat self-serving, but he did articulate the growing need for
scientific surveys independent of military or intelligence activities.
The state survey of New York was not the first, but during the
1830s it created some of the most competitively sought jobs for
naturalists and established important legal precedents for natural
history collections at just the time Congress was debating the
interpretation of James Smithson's bequest.[35]

2 The Lessons of Nature

The idea of all creation as a Book of Nature has long played a central role within the development of natural history. Despite the absence of any heuristic value, this humble approach to nature as a book whose pages are open to all has remained one of the most successful and certainly one of the most attractive ideas in the history of science. Charles Willson Peale found the inherent democratic values so appealing that one of his early tickets (equal to twenty-five cents) shows nature as an open book with the inscription:

> The Birds and Beasts will teach thee!
> Admit the Bearer to PEALE'S MUSEUM,
> Containing the Wonderful works of NATURE!
> and Curios works of Art.[1]

In the United States, the development of natural history museums followed the progress of two sciences in particular, Jefferson's study of large fossil animals (paleontology) and Wilson's study of birds (ornithology). Indeed, for more than a century, the discovery and accumulation of exotic birds and giant bones, later combined with the Victorian taste for big game hunting, characterized North American museum history. In the United States, public museums preceded the large private collections typical of

European trends, and in 1773, the Charleston Library Society opened the first American museum devoted to natural history. Peale's more ambitious Philadelphia Museum followed in 1786 and became the first commercial gallery to use overhead lighting. Located in the Philosophical Hall of the American Philosophical Society from 1794 to 1802, Peale's museum eventually exhibited more than 243 birds and 212 mounted quadrupeds, as well as fish, shells, rocks, and insects.

During the late eighteenth century, the grand voyages of circumnavigation launched by France, Britain, and Holland created the foreign study collections that gave rise to the science of ornithology. The National Museum in Paris, perhaps the finest in the world, contained only 463 bird specimens in 1793, and although that number increased by 3,000 within the decade, neither the brandy nor the embalming herbs used to transport and store the specimens warded off destructive insects after the bird skins were stuffed.[2] Peale surmounted many of these preservation problems and worked out basic methods for taxidermy and exhibit dioramas which had long-term consequences for museums as permanent scientific repositories of public interest. Apprenticed as a saddlemaker in his youth, Peale studied painting briefly in London. After the Revolution, he settled down in Philadelphia to the life of a portrait painter, until one summer day in 1784 a Hessian officer brought the artist some massive bones to sketch. Peale began to plan a natural history collection in conjunction with his portrait gallery and, with Thomas Jefferson's endorsement, undertook the management of the first public collections with national, rather than regional, emphasis.

Departing from European custom, Peale displayed bird specimens in glass-fronted cases backed by naturalistic views or habitats painted by his artistic children. His effective arsenic method for treating specimens was imitated safely by others, and using his early training as a saddlemaker, he stretched animal skins over wooden forms to restore lifelike attitudes. Often he hand-carved the internal limbs for these mounts and molded the glass eyes himself.[3] About fifty of Peale's mounted birds exist today at the Museum of Comparative Zoology in Cambridge, Massachusetts,

The Artist in his Museum, 1822. Charles Willson Peale (1741–1827) has
portrayed himself as a vigorous octogenarian in the famous Long
Room of the Philadelphia Museum. In the foreground are symbols of
his skills: a palette, the mastodon bones, and a taxidermy box with a
female wild turkey which his son Titian procured on the first Long
Expedition, 1819–20. Behind the raised curtain stand the glories of his
museum: the reconstructed "mammoth" and his bird collection with
the bald eagle drawn by Alexander Wilson in the upper left.
(Pennsylvania Academy of the Fine Arts)

where after almost two hundered years, the colorful Chinese pheasants Peale obtained from George Washington are still on display.[4]

By 1816, Peale's annual gross receipts of $11,924 indicated a paid attendance of nearly forty-eight thousand people, and Peale's sons attempted to start similar enterprises in Baltimore and New York, where Peale's efforts were being imitated unsuccessfully at Delacoste's museum and other less serious cabinets of curiosities. Many of Peale's visitors preferred his animal displays to his paintings, but Peale's artistic creativity was not easily suppressed, and he used his museum space for magic lantern shows and musical presentations.[5] His younger son, Titian, later claimed that his father had helped build the first organ in Philadelphia.[6] Peale used every means at his command to raise money for his museum, and the message of his musical "Dirge for Aldrovandus" was that the founder of the first museum in Bologna "spent his final days in the alms house."[7] Peale's undeniable triumph and the one that brought him a modicum of commercial gain was the exhumation of mammoth bones from a swamp in Orange County, New York. He mounted one entire skeleton in his museum in 1801, and with the aid of his sons he toured a second skeleton in Europe. Peale's mastodon, the first complete fossil reconstruction, made the mammoth a household word and established the sine qua non for natural history museums worldwide.

Regarding his own motivations, Peale concluded, "The gratification which every new object produced in the mind of an enthusiastic man is all powerful."[8] An important source of employment for the proprietor's many sons and daughters, former slaves, numerous in-laws, and the city's overflow of talented bank note engravers, Peale's imaginative displays instructed more than a generation of visitors before the collections were acquired by P. T. Barnum in 1849. In Peale's hands, good taxidermic methods permitted the safeguarding of valuable American type specimens. Specimens first given their scientific names by Alexander Wilson, Thomas Say, and Charles Lucien Bonaparte survived Barnum's sideshows before their final acquisition by the Museum of Comparative Zoology in Cambridge, Massachusetts. The objects of

Peale's broad attention raised the profile of American natural history, particularly zoology, and his bird collections permitted the production of Wilson's splendid bird book.[9] This landmark publication in turn inspired a group of Philadelphians to found the Academy of Natural Sciences in 1812 to "render ourselves as independent as possible of other countries and governments" and from the city's botanically oriented Linnaean Society.[10]

From its humble beginnings in John Speakman's apothecary shop, the Academy of Natural Sciences was organized around systematic zoology, and by the mid-nineteenth century, the academy's bird collections, first curated in 1817 by Peale's son Titian, were the largest in the world.[11] During the nineteenth century, other outstanding public institutions of natural history also focused upon the sciences of animals including man, and in contrast to the National Museum of the Smithsonian Institution or the American Museum of Natural History, no botanical collection emerged as a national herbarium even though American botany had existed as an intellectual pursuit before the publication of Linnaeus's *Systema naturae* in 1735.[12]

Peale's early displays reflected European ideas because despite his own nationalism, there were few authoritative American publications for him to use. Like his scientific contemporaries, the major authority Peale recognized was Carl von Linné, the brilliant Swedish biologist known to the Latin reading world as Linnaeus, and Peale organized his museum according to the Linnaean system, which divided nature into three kingdoms of minerals, plants, and animals.[13] These, in turn, were subdivided into classes, orders, genera, and species arranged into a single scheme. For animals, the Linnaean system ranged from the lowest, worms, to the highest, man. By the tenth edition of his *Systema naturae,* Linnaeus had assigned names to more than four thousand species.[14] Although Peale preferred native names for American species, he did adopt the Linnaean convention of Greek or Latin binomial nomenclature. Bartram's voluminous book of travels used Linnaean scientific names, but Peale's displays exposed the American public to Linnaean nomenclature in an easily understood fashion for the first time.

Linnaeus, like his predecessor in England, John Ray (1627–1705), had treated the individual species as integral and unchanging elements of nature. In 1762, he conceptualized the events of creation on a small island. First, the ordinal type was created, which, with time and interbreeding, gave rise to genera and species. Such a process gave biological reality or natural basis to the systematic divisions within the Linnaean framework. The order historically preceded the genus, and the generic type preceded the species. Although he could not offer a program of interbreeding that would permit the creation of a species from the generic type, Linnaeus did not believe this last step, the species, was accidental. Rather, Linnaeus believed that the origin of species was a final goal in the divine plan of creation. Linnaeus realized that knowledge of the true "system of nature" was "not within the immediate reach of human capacity," but he also recognized that "it is the exclusive property of man, to contemplate and to reason on the great book of nature."[15] This "book" was written in "the great alphabet of nature." These "elements of all science" were the key physical characteristics of plants and animals, "those marks imprinted on them by nature." For Linnaeus and every other naturalist who followed him, including Peale, it was a human duty "to affix to every object its proper name." Thus naming and the problems of nomenclature were integral aspects of the development of natural history.

Linnaeus died within a decade of Buffon, and Baron Georges Cuvier (1769–1832) quickly filled the position of authority they had enjoyed within natural philosophy. By the turn of the century, the Parisian zoologist had dramatically identified bones on both sides of the Atlantic as the remains of exotic animals whose past existence posed profound questions about nature and creation.[16] As one tourist remarked upon seeing Peale's mounted mastodon, "Perhaps we ought to imagine Noah found it too large and troublesome to put in the ark, and therefore left the poor animal to perish."[17] On a more serious level, these big bones also attracted the attention of Thomas Jefferson, and it is no accident that the early government expeditions he organized exercised stringent control over vertebrate collections even though the botanical materials were often more valuable to the medical profession and usually arrived in better condition.

Although Peale's Long Room was the prototype of the modern natural history museum, his museum idea was waylaid by two problems. The first he correctly identified as the need for reinterpretation of the Constitution so that federal funds could be appropriated to support national collections. The best of the Lewis and Clark collections went directly to Jefferson, who maintained them privately as part of his Poplar Forest estate, while Peale received such unwanted items as infested antelope skins or rowdy grizzly bear cubs. For the next two decades Peale's museum was used as a repository for the collections made on government expeditions to the West, but he was not reimbursed for his services. Charles Willson Peale died in 1827, and to Titian Peale's dismay, Congress did not give attention to the management of expedition collections until after 1842.[18]

A second obstacle was the location of type specimens, those irreplaceable biological specimens first given their scientific names and descriptions in press. By the time Peale's museum began to meet serious financial difficulties after his death, type specimens for North American species (with the notable exception of birds) were housed abroad or, in the case of invertebrates, dispersed across the country in private collections that were poorly curated and often destroyed by uninterested heirs. Peale's museum contributed to the intellectual climate that followed the War of 1812, which encouraged American studies and the growth of scientific societies and lyceums. These groups attempted to coalesce private holdings and in some cities almost achieved true museum status. In Philadelphia, the Academy of Natural Sciences, for example, opened its collections to the public in 1828, and by midcentury its holdings numbered 150,000 specimens.[19]

Throughout the eighteenth century scientific publications on American subjects had been controlled by the foreign press. After 1743, Mark Catesby's lavishly illustrated natural history of the American Southeast quickly became a collector's item, and the dearth of available scientific literature led Jacob Bigelow (who later coined the word "technology") to despair that what "we sorely lack in this land of Milk and molasses is good Botanical books and plates." "Upon the faith of a Mussulman," he continued, New Englanders "are as ignorant of botany as a nation of Jackasses."[20] In fact, during the eighteenth century, few persons in North America owned more than a Bible or Book of Common Prayer. New Englanders were not the only ones who were unread. American libraries were so few that on the Eastern Shore of Virginia, only three bookcases were recorded in wills before 1750, and by the Revolution even the city of Philadelphia could boast only one hundred cabinetmakers.[21] Peale's museum asserted the value of books and scientific publications.

The disagreement of foreign authorities such as Lamarck and Cuvier on many issues permitted Peale a refreshing latitude and freedom from dogma that contributed to the museum's initial popularity. Furthermore, Peale was genuinely interested in sharing the success of his natural philosophy. Between 1799 and 1802, the forty lectures he delivered summarized current knowledge about American birds and animals, and after 1820 he hired three of the city's most promising young naturalists, Thomas Say, William Keating, and Richard Harlan (1796–1843), to teach natural history, mineralogy, and medical anatomy at the museum.[22]

The museum offered enthusiastic naturalists specimens to study and draw and an audience eager to read and buy their published works of natural history. Specifically, Peale's various collections of animals and displays provided illustrative materials for the first volume of Say's *American Entomology* in 1824 and John Godman's *American Natural History* published two years later. Richard Harlan's *Fauna Americana* (1825) and Thomas Nuttall's two-volume bird manual of 1832–34 relied almost exclusively upon Peale's specimens.

In 1830, a writer for the *Cabinet of Natural History and American Rural Sports* remarked upon the value of natural history for all walks of life, including art, and claimed that the number of mammals Buffon ascribed to the New World, seventy, had been "much increased."[23] This increase in knowledge reflected activity at Peale's museum and demonstrated remarkable industry on the part of academy members during the fifteen years since the Peace of Ghent.

The name of the Academy of Natural Sciences revealed the members' lofty goals, but in the early years their attention was restricted to natural history. By 1842, the academy was strongest in the fields marked out in an earlier century by Linnaeus, Buffon, and Lamarck—botany, entomology, conchology, and ornithology. Linnaeus had not distinguished mollusks from echinoderms or worms, but, unlike Peale, academy naturalists were influenced by Lamarck's work on lower animals, and as a result, they recognized the inadequacy of the lowest Linnaean class, "vermes."[24] Rafinesque and Say prepared large, separate works devoted to mollusks, and Say with his characteristically humble patriotism introduced the American earthworm to the annals of science. In 1817, the founding members of the New-York Lyceum of Natural History drew up an early program that outlined such distinct research assignments as "helminthology," "polypology," "plaxology," and "apalogy." The objects of these elaborately designated sciences were worms, sponges, crustaceans, and mollusks respectively.[25]

More progress was made in some areas than in others, and the study of the larger cold-blooded animals—fish, amphibians, and reptiles—was uneven. The subject of the first zoological paper published in the United States was fish.[26] The author, William Dandridge Peck (1763–1822), was a New Englander who pioneered methods for mounting fish, and during the first quarter of the nineteenth century, the study of fish made more advances in the eastern states than in Pennsylvania. Indeed, an illustrated description of an important food fish by the nation's leading architect, Benjamin Henry Latrobe, in the major scientific journal of his day, the *Transactions of the American Philosophical Society,* was overlooked for almost two hundred years.[27] One reason for this oversight on the part of Latrobe's colleagues was that the nation's leading ichthyological resource was New York City's fish market. When Peale traveled to New York, he obtained specimens for his museum from the market, and Mitchill boasted in his encyclopedic study of New York State fish published in 1814 that sixty-two of the hundred new species he described were available in the city's public market.[28]

Mitchill was the founding editor of the *Medical Repository,* and it was he who encouraged Rafinesque to publish his American observations after his first visit to the United States in 1802.[29] When he returned to the United States in 1815, Rafinesque, like Mitchill, became a member of the New-York Lyceum. Never one to be modest, Rafinesque claimed after only five years in the field to have discovered more than 400 new fishes in midwestern waters. He also stated that by 1820 Charles Alexandre LeSueur had illustrated 150 species of North American lake fishes "on the plan of Wilson's ornithology."[30] Rafinesque usually held the accomplishments of other workers in the United States in low esteem: "Our Cat-fishes, eels, shads, sturgeons, &c, are for them mere fish to fill their stomach! and moreover, they are all of European breed, and were carried here by Noah's flood direct from the Thames, the Seine and the Rhine!"[31] Rafinesque was correct that many taxonomists were less eager than he to distinguish new species. As a result, his *Ichthyology of the Ohio River* has become a scientific classic, although, unfortunately, he did not cite LeSueur's unpublished work, which is now lost. Thirty years after Rafinesque's death, Louis Agassiz explained: "That he should have taken no notice of LeSueur's descriptions, is the natural consequence of the assumption upon which Rafinesque works throughout, that the fishes of our western waters differ uniformly as species from those of the Atlantic streams."[32] A veteran of the famous Péron Expedition of 1802, LeSueur impressed his Philadelphia colleagues with his considerable abilities. Throughout his American career, however, he was unable to procure an editor to translate his manuscripts from his native French, and he returned to Europe, where he died, having published virtually nothing of his extensive research.[33]

Western expeditions, even those attended by naturalists, had surprisingly little impact on ichthyology. The official catalog of species for the Stephen H. Long Expedition of 1819–20 does not name a single fish, although young Titian Peale drew specimens in watercolor and Say's manuscript notes discuss the perch and include a list of fishes copied from Rafinesque's book.[34] Long's second expedition in 1823 followed the major waterways of the northern Midwest. Again, although the report lists insects, mollusks, and even worms, fish are given only brief notice, the most exciting being the paddlefish.[35]

The study of reptiles made more headway, in part because the passages in Bartram's *Travels* which readers found most memorable included descriptions of the turtles and alligators in East Florida. Bartram wrote that the male alligators "bellow in the spring season" and that "vapor rises from their nostrils like smoke." Bartram also made numerous studies of snakes. The kindly botanist described the rattlesnake, which even the Indians feared, as "a wonderful creature, when we consider his form, nature and disposition."[36] Bartram's public remained more credulous than convinced. When Latrobe drew a detailed dissection of a rattlesnake for Benjamin Smith Barton in 1799, the eminent physician encountered so much superstition that he had to address the supposed "fascinating power" of the snake in the leading scientific publication of the time. Barton shrewdly realized that his would not be the last word, for "he who seriously believes will not stop here."[37] Thirty years later, the public knew little more about rattlesnakes. For example, plate XXI of Audubon's *Birds of America,* which showed a rattlesnake robbing a nest of mockingbirds, was labeled "Drawn from Nature." Rattlesnakes do not climb trees, and Wilson's editor, George Ord, insisted that the plate was "a miserable fabrication" and "one of the clumsiest lies ever invented."[38] Audubon stood by his claim that he had actually observed the "habit of climbing in the Rattlesnake" and probably benefited from the publicity that accompanied the ensuing controversy in the scientific press.[39] Despite public interest in snakes, Ord's co-workers were to some degree responsible for the neglect of definitive herpetology. On the first Long Expedition, Say did not catalog the six new species of rattlesnakes or the new toad and two new lizards he discovered. Perhaps the entomologist hoped that Harlan, who described the salamander *Triton lateralis* for him, would work up descriptions for the other new western reptiles.[40] Say did record "a large and beautiful animal" he named *Ameiva tessellata* for its mosaiclike qualities, and LeSueur described the map turtle, *Testuda geographica,* but other reptiles of the well-traveled Mississippi and Missouri river basins remained poorly known before midcentury.

Pickled in whiskey, reptile collections were relatively easy to transport in barrels over long distances. In contrast, mammalian specimens posed formidable problems in an era before refrigeration, and classification of western mammals, in particular, was delayed by an unexpected aspect of the Lewis and Clark Expedition. The government had not created an organization to receive the expedition's collections.[41] Furthermore, the materials Jefferson forwarded to Peale for preservation were hardly a zoological bonanza: "The uncleaned bones bred the Insects which afterwards fed on the Skins."[42] Peale hoped to illustrate the published natural history of the expedition and made a few sketches, but potential authors, proceeding cautiously during the period of Jefferson's political unpopularity, did not want to preempt the discoveries of the popular explorers by publishing before they did. This situation was further complicated in 1809 by the bizarre death of Meriwether Lewis.[43] William Clark attempted to find a second author, but Barton, who agreed to undertake the publication at Jefferson's urging, died in 1815 having done very little toward its completion.[44]

Barton's death was probably more detrimental to an early American publication on mammals than historians have realized.[45] The last of the fourteen species Ord attributed to Lewis and Clark was not named until 1857, a full half century after the expedition's return.[46] All the same, western expeditions did give impetus to American mammalogy. The Long Expedition of 1819 retraced parts of the earlier Lewis and Clark route to collect additional specimens for Peale's museum. Say described eight new western mammals, and better specimens enabled Ord to identify one of the Lewis and Clark species.[47] En route to Oregon in 1834, Nuttall and John Kirk Townsend found a new mole and a chipmunk, both of which now bear the latter's name and were figured in Audubon's *Viviparous Quadrupeds of North America*. Of course, not all new mammals were found in the West. The Philadelphia anatomist Harlan described three new bats in 1831 and a new mouse Nuttall gave him from Norfolk, Virginia.[48]

Audubon claimed that Nuttall, despite his years in the field, could neither swim nor shoot, and it is true that Nuttall like Wilson kept a number of species as pets to study.[49] Peale's uncanny ability to tame and even to train such feral creatures as the bald eagle and the hyena also did much to encourage an almost anthropomorphic approach to captive mammals while the museum was located at the State House.[50] The character of the museum exhibits changed, however, as the aging proprietor relinquished management to his sons, and Peale's youngest son, Titian, the exuberant marksman who had assisted Say out west, mounted the new mammals with snarling faces.[51] In an effort to salvage a severely damaged specimen of the then novel mule deer, Titian detached the head and displayed it being savaged by a prairie wolf. Such realistic mounts must have been memorable, but mammalogy, even at the museum, was overshadowed by vertebrate paleontology. Nuttall's description of the typical mammal, "large head, formidable jaws armed with teeth, the capacious chest, wide shoulders, and muscular legs," suggests that despite his extensive western travel, the American quadruped the English botanist found most impressive was the mammoth reconstructed in Peale's museum.[52]

The discovery of large fossil forms had made a great impact upon American ideas. As William Cooper (1797–1864) remarked, in every case—*Mastodon, Megalonyx, Elephans, Bos,* or *Cervus*—these fossil giants were recognized as both unique to North America and extinct.[53] Ironically, many of these fossil types became better known during the early nineteenth century than the living species with which they were compared. In 1826, James Ellsworth DeKay, president of the New-York Lyceum, complained that the best-known North American quadruped, the white-tailed deer, *Cervus virginianus,* had been described by a foreigner, and DeKay might have added that, despite Titian Peale's imaginative mount, the mule deer was still undescribed.[54] Furthermore, the grizzly bear, which Ord had designated in 1815 as *Ursus horribilis,* was so poorly known in the wild that DeWitt Clinton proposed it as the source for the great claws which Jefferson received from John Stuart's saltpeter mines in western Virginia.[55] The fossil claws proved to be those from a giant ground sloth *Megalonyx.*[56] Jefferson initially believed that the "great claw" was a carnivore, a giant cat, as "formidable an antagonist to the mammoth as the lion to the elephant."[57] The great size of the mammoth prompted similar responses among viewers at Peale's museum. After Rembrandt Peale inverted the tusks to resemble scythelike fangs, one visitor marveled "that such a huge carnivorous monster should have even existed" and wondered nervously, "has it now ceased to exist?"[58] Jefferson continued to argue that the mammoth did exist. Although he realized it was herbivorous, he was hard put to explain the plant diet of a mastodon found in the frozen deserts of Siberia. In contrast, Clinton, with the equivocal wisdom of a veteran career politician, hazarded that the creature was both a carnivore and an herbivore.[59] Clearly, American mammalogy had not advanced to the point that these conclusions were mutually exclusive. Although Cuvier had already identified most of these fossil finds, his brilliant correlations of anatomy, dentition, and diet had not yet become imperative for American naturalists, and an edition of his work—his geology—was not published in the United States until 1818.[60]

Jefferson's thesis that large quadrupeds retreated before European man onto the Great Plains was reasonable, but none of his scientific contemporaries, not even admirers like Clinton or Mitchill, shared his "benevolent persuasion" that mammoths awaited discovery in the continental interior.[61] Edwin James (1797–1861), a young botanist from New York State, spoofed Jefferson's unlikely expectation for military science when he recalled elk spotted for the first time during the spring season of the first Long Expedition: "The effect of *mirage,* together with our indefinite idea of distance, magnified these animals to a most prodigious size. For a moment we thought he was the mastodon of America, moving in those vast plains, which seem to have been erected for his dwelling place."[62] Mammoths aside, neither James nor his colleagues observed that other "terror of all western travelers," the grizzly bear, and James had to prepare his official account of the expedition from hunters' accounts, Clinton's paper, and mounts (now more than two decades old) at Peale's museum.[63] In 1820, James did witness the escape of a grizzly chained in the yard of the Missouri Fur Company at Council Bluffs. The more seasoned men fled, but James innocently stood his ground as the bear "in his round came to me" and "rearing up, placed his paws on my breast." Wishing to rid himself of "so rough a play-fellow," James, with a foolhardy boldness gained perhaps from Peale's Philadelphia menagerie, turned him around, and the bear "ran down the bank of the river" and, to everyone's relief, took a bath.[64]

Jefferson was correct in realizing that the American interior would provide a large observatory for the relationship of extermination to the question of extinction. As state senator, Mitchill supported a bill "for the destruction of wild beasts and the encouragement of history."[65] Less political naturalists were distressed by the "wanton destruction" of biological species before their natural histories were understood. In 1819 Say, aware that many large species were disappearing, chose to concentrate upon the Indians instead of his zoological assignments, and he spent the winter months recording their languages, customs, and material culture. By 1832, the painter George Catlin suggested that the region of the Great Plains be made into a "preserve" for both the Indian tribes and the "fleeting herds" in "all the wild freshness of their nature's beauty."[66] Catlin had visited Peale's museum and later toured his own Indian gallery in Europe to make their plight better known among "refined citizens and the world." The lesson of nature for other explorers who made western collections for the Philadelphia Museum and the academy was that eradication of large game mammals would lead to the end of the Indian groups dependent upon them. American natural history, they would argue, required a broader vision than the Linnaean method or museum studies.[67]

Drawn from Nature

3

Despite his poverty, the ornithologist Alexander Wilson embodied the Romantic ideal of genteel hero exploring the American wilderness with a higher vision of nature.[1] One of his reviewers speculated that had Wilson lived beyond 1813, his "original genius" and his opportunities to wander might have made him "like Wordsworth's Peddlar, a good Moral Philosopher."[2] Only one year after Wilson's death, Wordsworth, an admirer of Bartram's *Travels,* may well have had Bartram's protégé in mind when he composed "The Excursion," the poem in which the Peddlar appeared. In any event, there can be little doubt that the enthusiastic reception of Bartram's *Travels* by the Lake Poets did much to condition readers' expectations for Wilson's career.[3]

In 1808, the first volume of Wilson's beautifully illustrated *American Ornithology* was the most ambitious publication undertaken in the United States. Wilson's work united the considerable local intellectual resources of the Philadelphia area in an enduring accomplishment. In its final form, the nine-volume *American Ornithology* stands as a tribute to scientific figures of the American Enlightenment then in their senior years—William Bartram, Benjamin Smith Barton, Charles Willson Peale, and Thomas Jefferson—as well as Wilson's posthumous editor, George Ord. Wilson wisely took advantage of Bartram's retirement in Kingsessing to avail himself of the aging naturalist's ornithological knowledge.[4] Wilson depended upon Peale's goodwill for everything from bird skins and mounted specimens to labor and art supplies. Knowing that Peale himself had hoped to illustrate the birds sent back by Lewis and Clark, Wilson was so careful in his annotations that his volumes provide more details about Peale's bird collections than any other single source.[5] As a result, seventy-one of the eighty-five birds Wilson described can be associated with Peale's cataloged specimens. Wilson hoped to promote amateur interest in ornithology, and methods he recommended to his readers were based upon Peale's taxidermic techniques, without which Wilson, with his technical limitations, could not have proceeded:

As soon as the bird is shot, let *memoranda* be taken of the length, the breadth (measuring from tip to tip of the expanded wings), color of the eyes, bills, legs and feet, and such particulars of its manners, &c, as may be known. Make a longitudinal incision under the wing, sufficiently large to admit the body to be taken out; disjoint the wing close to the body under the skin, and endeavor with a pair of scissors or penknife to reach the neck, which cut off; pass the skin carefully over to the other wing, which also disjoint and separate from the body, then over the whole body and thighs, which last cut off close to the knees; lastly, separate the whole skin from the body at the roots of the tail feathers, which must not be injured. Return to the neck and carefully pass the skin to, and beyond, the eyes, which scoop out; cut off the neck close to the scull, penetrate this way with your knife into the brain, which scrape completely out; dissect all the fleshy parts from the head, wings, and skin; rub the whole inside with a solution of arsenic, sprinkle some of the same into the cavity of the brain, throat, &c.; stuff the vacuity of the brain and eyes with cotton, to their full dimensions; return the skin carefully back, arranging the eye lids and plumage; stuff the whole with cotton to its proper size and form, sew up the longitudinal incision, and, having carefully arranged the whole plumage, sprinkle it outwardly with a little powdered arsenic; place it in a close box, into which some camphor has been put, and cover it with cotton or ground tobacco. In the whole operation the greatest care must be taken not to soil the plumage with blood.

If arsenic cannot conveniently be had, common salt may be substituted.[6]

Wilson also relied upon the technical expertise of Alexander Lawson (1773–1846), a fellow Scotsman and valued friend, who

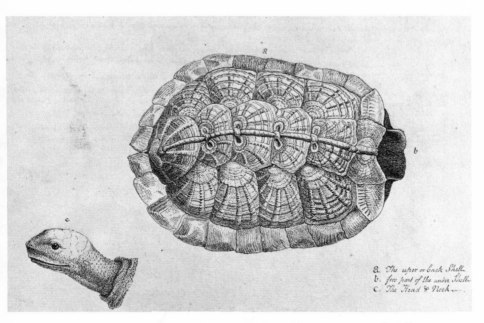

a. *The upper or back Shell*
b. *fore part of the under Shell*
c. *The Head & Neck*

Testudo calata. This sketch of the wood tortoise is not one of William Bartram's best drawings, but it is interesting because the head and obviously the legs and tail have been dissected from the shell in marked contrast to the scientific illustrations of his followers. (American Philosophical Society)

was probably the most proficient engraver at work in the United States. Four years before the first volume of the *American Ornithology* materialized, Wilson wrote to Lawson for artistic guidance.[7] He realized that his plan to illustrate birds in a city where most professional artists developed their skills engraving bank notes must have seemed quixotic to this thrifty Scot. Yet Lawson, at no profit to himself, remained loyal to the project long after Wilson's death, and with Ord's financial assistance, engraved both the reprinted and supplemental volumes of the *American Ornithology*. The statement "Drawn from Nature" that Lawson had printed at the bottom of each illustration masks the exact roles of the discovering naturalist, the living species, the museum specimen, and the illustrator. Like Audubon, Wilson drew his birds in watercolor, ink, or pencil from mounted specimens. Lawson often

1. Louisiana Tanager. 2. Clark's Crow. 3. Lewis's Woodpecker.

20

Louisiana Tanager (1), *Clark's Crow* (2), and *Lewis's Woodpecker* (3),
1811. Engraved by Alexander Lawson from drawings by Alexander
Wilson, this plate was hand-colored for the *American Ornithology,*
possibly with pigments ground by Peale and his sons. The specimens
were obtained by Lewis and Clark and deposited with Peale for
preservation and mounting.

finished these sketches with landscapes, possibly copies of Peale's exhibit backdrops rather than original contributions. Other plates are simply composite compositions Lawson made using cutouts from several drawings by Wilson. The term "Nature" is thus somewhat misleading in Wilson's book because the background landscapes and the associations of species illustrated may or may not correspond to actual habitats or the place where the birds were found.

The mainstay of Wilson's success was plates "Drawn from Nature." The major colorist was Alexander Rider, a Swiss painter who later illustrated three volumes on mammals in Peale's museum. Rider was coloring Wilson's work from freshly mounted bird specimens, so he must have been associated with the museum as early as 1808. His early training in oils has often been blamed for the opaque treatment of some of Wilson's plates, but in some cases the fault lay with homemade pigments, one consequence of Jefferson's embargo. Wilson, for example, noted a yellow pigment "from the laboratory of Messrs. Peale and Son, of the Museum," which may have been used for the bright yellow neck and wings of the Louisiana tanager (now the western tanager) shown in plate 20 of his third volume (1810).[8] This new species had been forwarded to Peale by Lewis and Clark, and the fact that Wilson drew his bird from the admittedly "frail remains" of three specimens Peale had cleverly pieced together may explain why the bird of his plate lacks the feet so clearly drawn in Peale's earlier illustration of 1806. At least one of the female members of Peale's large family also contributed to the *American Ornithology*. Wilson's ledger notes show that in the fall of 1810, he paid Anna G. Peale $51.27 for coloring 254 impressions for the first two volumes of his work. Wilson paid another woman $78.18 for coloring 331 impressions of plates, and for volume 6 he spent $63.63 for nine plates of copper and $80 for the engraving, bringing the total cost for the coloring and lettering to $512.38.[9]

The grand size, colorful plates, and gracefully written prose of Wilson's monumental bird book opened a new era in American ornithology. In nine volumes (one completed after Wilson's death by George Ord), Wilson figured 320 North American birds representing 262 species. Of these, only 39 were new to science, but 23

Alexander Wilson (1766–1813), ca. 1809–13. Despite Wilson's poverty, this oil portrait by Rembrandt Peale shows Bartram's student, after his success in ornithology, as a "scientific gentleman" wearing a ruffled shirt. (American Philosophical Society)

others were described with sufficient detail to differentiate them from European species with which they had been confused. Wilson's accomplishment was achieved at great personal expense for he was marginally employed as a schoolteacher. In his book he had urged, "Let but the generous hand of patriotism be stretched forth to assist and cherish the rising arts and literature of our country, and both will most assuredly, and that at no remote period, shoot forth, increase and flourish with a vigor, a splendor and usefulness inferior to no other on earth."[10] Wilson charged $120 for subscriptions for his bird book, and despite his high hopes for "the generous hand of patriotism," the initial public reaction was often disappointing. He wrote with obvious reference to Peale's famous displays that in Newark the book "attracted as many starers as a Bear or Mammoth would have done." These, Wilson confessed, he would have "willingly exchanged for a few simple *subscriptions.*"[11]

Initially, only two hundred copies of volume 1 were printed. Wilson eventually procured about 650 subscribers, including Thomas Jefferson. The state of Pennsylvania topped the list with 113 subscribers, but George Ord noted with some cynicism that in Philadelphia, "of all her literati, her men of benevolence, taste and riches, SEVENTY only, to the period of the author's decease, had the liberality to countenance him by a subscription, more than half of whom were *tradesmen, artists,* and those of the middle class of society." In contrast, he obtained 60 subscriptions from "the little city of New Orleans."[12]

Wilson used every means at his command to broaden the appeal of his work. A well-known poet and political activist in his native Scotland, he put his verse to another use in the volumes of his great bird book, the cause of conservation. A six-stanza poem, for example, pays sentimental tribute to the "rural innocence" and usefulness of the bluebird:

> He drags the vile grub from the corn he devours;
> The worms from their webs where they riot and welter;
> His song and his services freely are ours,
> And all that he asks is, in summer a shelter.[13]

Wilson was defending bluebirds against the guns of farmers, whose ignorant and excessive destruction deeply alarmed him, for he considered ornithological species a national treasure. Although Wilson never procured the number of subscribers his book merited, the *American Ornithology* became a compelling vehicle for learning and nationalism, and the response in the press was extremely favorable. Samuel Mitchill commented: "The exquisite paper, the distinct type, the correct engraving, and the fine colouring are all domestic; and the same American character belongs to the presswork, the binding, and the other mechanical parts of the work."[14] In 1814, a reviewer for the *Port Folio* agreed: "The magnificent performance, so honorable to the country, and so eminently calculated to shed a lustre on the era which has produced it, has reached in its progress its eighth volume."[15] Other critics were impressed that a poet without means could produce work comparable to that of the best European ornithologists. They hailed his life as "the complete triumph of genius over want of education, and of persevering industry over the evils of poverty."[16] Another reviewer observed that while toiling naturalists "make efforts for future and perhaps posthumous renown," they were "obliged literally to do present work for nothing, and procure meat, fire, and clothes from some other source."[17]

Despite his considerable personal sacrifices, Wilson made American natural history available in elegant form for the first time, and as Governor Clinton hoped, underscored the need for more scientific publications in the United States.[18] Wilson's timing was certainly critical to his success. In 1809 the first geological map of the United States was published as the result of the largest survey of its kind in the world.[19] American science was becoming better by being bigger. Within a decade, the New-York Lyceum of Natural History was established, and Benjamin Silliman founded the influential *American Journal of Science* in New Haven.[20] Not surprisingly, Wilson's example was most significant in the city where his books were published, and in 1812 a group of young naturalists pledged "to bring forth the resource that God has given us within ourselves to enable us to withstand the deprivation of foreign trade and render ourselves as independent as possible of other countries and governments."[21] Wilson's death in 1813 at the height of his career made apparent the need for organized scientific endeavor free from the vacillations of royal patronage. His success was a key factor in the decisions of foreign-born naturalists Thomas Nuttall, C. S. Rafinesque, William Maclure, and C. A. LeSueur to work in America, and their individualistic approaches to financial support for science corroborated the professional needs which Wilson's publications first made apparent.

Thomas Jefferson had suggested in his *Notes on the State of Virginia* that North American faunal species were distinct forms independent of European types. In 1789, he had written the president of Harvard College from Paris that he did not find a single land bird common to Europe and the United States, and he questioned "if there be a single [common] species of quadrupeds."[22] Fourteen years later, Jefferson persisted in this belief, which had momentous appeal to a younger generation of naturalists.[23] He wrote to Alexander Wilson on April 7, 1805: "I am of opinion there is not in our continent a single bird or quadruped which is not sufficiently unlike all the members of its family there to be considered as specifically different."[24] Wilson admired Jefferson as a statesman and a friend of science. There was a copy of Jefferson's *Notes* in the Bartram family home, where Wilson was a frequent guest, and if he had not read the work before 1805, he soon became aware of the author's thesis. On Bartram's recommendation, Wilson and Jefferson became correspondents and discussed ornithology together in the White House on December 17, 1808.[25]

Throughout the *American Ornithology,* Wilson's vision of the freedom available within the natural world of the new nation reflects a poetic interpretation of Jefferson's ideas. Wilson's volumes are a long defense of the quality and validity of North American avian species. For example, he reminded his readers that unlike the "whimsical philosopher" Buffon, he felt compelled to allot "to each respective species" that "rank and place in the great order of nature to which it is entitled." His description of the golden-winged woodpecker (now the yellow-shafted flicker Linnaeus thought was a cuckoo) denounced "the abject and degraded character which the count de Buffon with equal eloquence and absurdity, has drawn of the whole the tribe of woodpeckers." Like Bartram, Wilson reexamined the rationale for class structure and slavery in the theory of degeneration. As he discussed the feeding habits of the diminutive downy woodpecker, the smallest American woodpecker, he renounced what Buffon had called "incessant toil and slavery," "painful posture," and "a dull and insipid existence." These he claimed were expressions "improper" to natural history because they were "untrue and absurd."[26]

Wilson was not without humor. Comparing the hairy woodpecker to the smaller downy woodpecker, he queried, "I wonder how it passed thro the count de Buffon's hands without being branded as a spurious race, degenerated by the influence of food, climate or some unknown cause." Wilson elaborated this theme in another place: "This eternal reference of every animal of the new world to that of the old, if adopted to the extent of this writer Buffon, with all the transmutations it is supposed to have produced, would leave us in doubt whether even the *Ka-te-dids* of America were not originally Nightingales of the old world degenerated by the inferiority of the food and climate of this upstart continent." To the contrary, Wilson exhorted his readers: "Let us examine better into the operations of nature, and many of our mistaken opinions, and groundless prejudices will be abandoned for a more just, enlarged and humane mode of thinking."[27]

After Wilson's death, subsequent authors at the academy and many of their readers continued to argue that American works sold by subscription could not only successfully rival European publications but also repay "with interest the debt which America owed Europe."[28] Wilson's persuasive approach had worked so well on the pages of the *American Ornithology* that, following the War of 1812, Americans no longer considered a refutation of Buffon's theory of degeneration a necessary preamble to a work of natural history. In 1815, for example, George Ord declared in the first comprehensive discussion of North American mammals: "It gives us pleasure to learn that the dogmata of this vain and whimsical philosopher have lost much of that regard which an imposing name has contributed to attract."[29] In 1828, Charles Lucien Bonaparte claimed that Buffon's theory was too absurd to merit refutation.[30] All the same, later editions and supplements of Wilson's popular work continued to give life to the idea that New World fauna were distinct species. This notion, initially a philosophical suggestion on Jefferson's part, became an intellectual cornerstone after the War of 1812. Wilson's admirers asserted that American species should be studied by Americans and that resulting collections should be housed in the United States rather than abroad.[31] Thomas Say, always discouraged by the lack of reference libraries, expressed a common sentiment to a colleague in Boston when he insisted: "We must therefore be content to labor onward as we may, with our efforts directed to the honour and support of *American* science, with *pro nobis* for our motto, not so much in a *personal* as a *national* sense."[32]

 For naturalists who relied upon Peale's museum collections, the message the "birds and beasts" taught was self-sufficiency and egalitarianism. Although the "laws of nature and of nature's god" were elusive, the academy's declaration of scientific independence in 1812 met with initial success. Within a generation, however, the limitations of this intensely nationalistic approach to natural history were evident. After 1842, "Drawn from Nature" did not necessarily imply good science, and naturalists who modeled their careers upon Wilson's no longer represented the vanguard of taxonomists in the United States. Wilson and Peale had used natural history to promote their favorite arts—poetry and painting—as well as the sciences, but the changing reception of illustrated technical works that followed their pioneering efforts demonstrates the close and eventually conflicting roles of artistic statement and scientific novelty in organized natural history.

PART TWO

Proposals

Writing the Book of Nature 4

The size and number of illustrated books of natural history published in Philadelphia following the War of 1812 reflected new collections made on western expeditions and an early commitment to the establishment of a distinctly American science. Naturalists took special pride in the belief that species could be sufficiently described using resources available in the United States: "We enjoy the most pleasing prospect," Richard Harlan boasted, "in the present rapid progress towards a complete illustration of the natural history of our country."[1] Yet despite the enthusiasm generated by such "rapid progress," fear of preemption by foreign authors persisted after the description of the most famous western collection, the Lewis and Clark herbarium, was published in London in 1814.[2] Not surprisingly, the need for prompt publication of new materials became especially real to those academy naturalists who had endured the hardships of western exploration—notably, Thomas Nuttall, Thomas Say, and Titian Ramsay Peale. For Say, publication in this country was not only a professional precaution but also a patriotic duty. On hearing that a fine collection of American Lepidoptera was to be sent to France, he implored a colleague to describe "this beautiful order & let them not be wrested from their own countrymen, thus unnaturally."[3]

Wilson's American readers anticipated good paper, handsome

55

typeface, and large colored plates[4]—requirements the membership of the Philadelphia academy was uniquely qualified to meet. Titian Peale and C. A. LeSueur were artists; Wilson, Nuttall, and Godman had been apprenticed as printers.[5] Thus Say and other members of the academy could avail themselves of colleagues' firsthand knowledge to continue the pictorial style of Wilson's bird plates.[6] For the next three decades, the experienced engraver, Alexander Lawson, and his colorists—Alexander Rider, Helen and Malvina Lawson—as well as James and Titian Peale, LeSueur, and Cornelius Tiebout (1777–1832) complemented the nation's natural history with thousands of hand-colored plates. Because academy naturalists often used the same publishers in Philadelphia, their books also shared a common character which readers quickly recognized.[7] Wilson's publisher was Samuel Bradford, for whom he had worked as assistant editor of a new American edition of *Rees's Cyclopaedia*. Later naturalists used Samuel Augustus Mitchell or Henry Carey and Isaac Lea.

Besides the valuable contacts between potential authors, artists, engravers, editors, booksellers, and subscribers which academy membership provided after 1817, the organization encouraged scientific cooperation and close working relationships. Say and Godman helped Bonaparte make idiomatic the English text for his supplement to Wilson's volumes that described materials Say and Titian Peale had collected in the West. Bonaparte, in turn, later financed Peale's collecting trip in Florida and assisted Godman's publication.[8] Nuttall furnished Say with about one-tenth of the western insects Say described in book form, and Say offered Nuttall the plants he had collected on the second Long Expedition to the source of the Minnesota River in 1823. Cooperation led to an identifiable systematic approach, but without the generous patronage of William Maclure, it is unlikely that these friendships would have proved so fruitful for science.[9]

Maclure was a wealthy businessman who had established his scientific reputation by undertaking the first geological survey of the United States. This work was first published in 1809, revised in 1817, and republished in 1818 with a detailed map.[10] On retirement from the field, Maclure became an ardent advocate of the scientific education of the poor, and he envisioned the United States as a place where (perhaps for the first time in history) public knowledge would actually reflect the progress of scientific research.[11] As the academy's second president, a position he held despite prolonged absences from Philadelphia from December 1817 until his death in 1840, Maclure funded field trips for LeSueur, Say, Peale, and George Ord. He established a valuable reference library (particularly rich in French works) in buildings he helped finance, and as a measure of his own patriotism, he bought copperplates engraved for foreign works on American subjects for reuse in cheaper American editions.[12] In 1817, he procured a printing press so that the academy members could organize an independent journal in which Say, Nuttall, LeSueur, and Ord published technical articles on new genera and species.[13] Yet despite the advantages of speed, cost, and larger editions, this inexpensive mode of publication did not seem to satisfy the ambitions of these authors, and they continued to publish and republish their findings as large books.

Within a period of twenty-five years, these authors created an illustrated library of American natural history. In 1817, Say issued the first part of his *American Entomology*, which was followed by three volumes from 1824 to 1828.[14] After publishing some of the first American lithographs in the academy's *Journal*, LeSueur began work on illustrations for his proposed work on American ichthyology.[15] Ord's 1824 edition of Wilson's volumes was accompanied by Bonaparte's supplements in 1825, 1828, and 1833. Godman's illustrated *American Natural History* of 1826 followed Harlan's more technical *Fauna Americana*.[16] In 1839, large-scale academy publications on American subjects continued with Samuel G. Morton's *Crania Americana* and Nuttall's *North American Sylva*.[17] These academy taxonomists evidently enjoyed the literary process of classification which the terse format of journal articles could not accommodate. Say, for one, delighted in Linnaeus's allusions to classical literature: "In pursuance of his attempt to unite natural and civil history, Linné divided his vast genus Papilio into several families, of which he named the first Equites or Knights. This family, containing some of the largest and most beautiful of the insect tribes, was subdivided into Trojans and Greeks. The former were distinguished by red spots on the breast." With characteristic Quaker aversion to the blows of war, Say went on to discuss the possibility of insects' experiencing the sensation of pain.[18]

Influenced by Maclure's social concerns, Say felt responsible to a reading public that financially supported his appointment to government expeditions.[19] In Edwin James's official report of the first Long Expedition, Say confined technical descriptions to the footnotes, in which, for example, a large *Lampyris* was perfunctorily described as "readily distinguished from the small species [the common firefly] by its mode of coruscating."[20] By contrast, in the *American Entomology,* the "mode of coruscating" became "Scintillations on a summer's evening . . . scarcely less abundant than the lights of the firmament, which they feebly, and but for a moment, rival."[21] Say prepared glossaries to his books explaining taxonomic language, and he was not unique in his efforts to make science accessible to new audiences.[22] Nuttall's *Genera* of 1818 was a technical catalog, but he enriched his illustrated supplement to Michaux's *Sylva* with accounts of his western travels. Like Say, Nuttall employed fanciful allusions. One genus found in the Blue Mountains of Oregon seemed to derive its changeable qualities from the aqueous element it loved. He recalled: "For many miles we never lost sight of the long-leaved Willow, which seemed to dispute the domain of the sweeping flood, fringing the banks of the streams . . . at every instant, when touched by the breeze, displaying the contrasted surface of their leaves; above of a deep and lucid green, beneath the bluish-white of silver; the whole scene, reflected by the water and in constant motion."[23]

Wilson had included entire pages of verse in his bird book, but Nuttall (like Say) never permitted such diversions to cloud his critical eye. When he compared the California sycamore to the eastern plane tree, he realized that "he sojourned in a new region of vegetation," and the objects "apparently the most familiar," associate them as he would, "were still wholly strangers."[24] His style satisfied American readers, although Ord did complain that Say's classical embellishment was too sparse.[25] Even in the closing decades of the nineteenth century, new editions of works by Say and Nuttall continued to provide an appealing alternative to the lurid view of nature "red in tooth and claw" celebrated on the popular level by American social Darwinists.[26] A more profound aspect of these large works received no critical notice. In their prefaces and essays, academy botanists and zoologists alike stated the need for reorganization of American taxonomic groups and extensive revision of nomenclature. Charles Lucien Bonaparte repeatedly noted taxonomic "confusion" and emphasized the need for extreme generic reforms.[27] Needless to say, academy naturalists were neither the first nor the only American scientists to recognize the failures of the Linnaean system.[28] By 1820, their geological colleagues frankly admitted that the natural history method of Linnaeus was no longer useful for the minerals because "from one extreme of the series to the other, there is a constant transition of approximating aggregates into each other."[29] Naturalists who had accompanied expeditions and whose results were included in the same reports could not overlook the implications of such geological considerations in their own work.

After 1817, the improved European microscopes demonstrated the inadequacies of referring all invertebrates except insects to the Linnaean class Vermes, or "worms."[30] At the academy a vigorous interest developed in the systematics of the lower forms Linnaeus ignored. Unlike their precocious contemporary at West Point, Jacob Whitman Bailey (1811–57), academy microscopists were not outstanding anatomists. Maclure had given the academy its first compound microscope and a "square" or single-lens microscope in 1817, but the record suggests that the novel instrument was used more as an aid to identification than as a research tool.[31] The microscope became part of the equipment taken on western expeditions, but chromatic aberrations handicapped its practical use in the field, and by the mid-1830s large crowds paid to view one of Rubens Peale's microscopes as a novelty at his New York museum.[32] In the laboratory, however, microscopical investigations presented puzzling questions of form and function, which American authors with their distrust of philosophy were not likely to solve. About mollusks, John Godman wondered "why so much elaborateness of construction, and such exquisite ornament. . . . Destined to pass their lives in and under the mud, possessed of no sense that we are acquainted with, except that of touch, what purpose can ornament serve in them? . . . We cannot suppose that the individuals have any power of admiring each other."[33] In other academy writing on invertebrates, a similar preoccupation with "exquisite ornament" and color prevailed over internal anatomy. As Titian Peale's description of a chrysalis shows, the study of invertebrates was indeed well suited to the artistic imagination, "for by one of those wise provisions of nature, which so often are obvious to the student of her works, it has been decreed that the green and gold shall be sported only whilst the prevalence of verdure serves to guard the insect in the most helpless state, from analogy of colors."[34] Thus the microscope, which enhanced an appreciation of form, ironically encouraged recourse to superficial taxonomic characters at the expense of a more biological orientation, and an early opponent of the use of the microscope for classification criticized the malacology of Say and Isaac Lea precisely for its neglect of physiology.[35] In truth, what Say had said of conchology could have been applied to other academy studies as well: "By the beauties of the subject it treats of, it is adapted to recreate the senses."[36]

The best academy authors, however, were not irrevocably tied to artificial (primarily Linnaean) systems of classification by their aesthetic concerns. As an early historian remarked, the academy's shell collection was arranged according to the system of Lamarck, not Linnaeus, and during the 1830s it was rearranged to follow Cuvier.[37] Although Say had settled in 1819 for an admittedly "superficial" arrangement of mollusks "taken from shells alone," he was quick to assert that he did not subscribe to "the inviolability of the Linnaean system." His lovely volumes of *American Entomology* reflect his sincerity in this regard and also reveal the ambiguity of his taxonomic designations. Each section of his text describes first the "generic type," which being ideal cannot be represented by an individual species or contained in a collection. Say argued that the true "characters" or features that defined a new genus could be determined for certain only as more new species of the genus were discovered. Since he often was working with one or at best a few members of a new genus, he chose characters that served primarily for identification. Coleoptera (beetles), the insect group that made up much of Say's collections, are highly variable, and Say, like his contemporaries, did not understand the biological significance of variation. Yet, being a keen observer, he could not minimize the troublesome phenomenon in other groups as well. In an effort to provide some "essential" characteristics for classifying *Acrydium ornatum,* Say said he "purposely avoided a minute detail of colours and markings, noting such only as will probably prove to be permanent, or nearly so, and characteristic of the species."[38]

In the absence of a larger reference collection, the concept of the ideal type provided Say with a needed taxonomic refuge. If the characters of a species were even less "permanent" than those of the flying grasshopper (the example above), Say simply placed the specific observations under a generic heading. His efforts to classify new species were also complicated by questions peculiar to the field naturalist. A western blister beetle Nuttall had given him exhibited the behavior of the genus *Lytta* but the antennal form of the genus *Mylabris*. In this case, Say emphasized function over form and retained the *Lytta* classification.[39] In sum, Say was not always consistent. He used what he considered to be the best foreign authorities and attempted to fit their identifications with his own. Apparently uncertain of his success, he left the pages of his volumes unnumbered so that the plates and accompanying text could be rebound to suit the subscriber's needs. Besides Say, there were other academy authors who were unwilling to endorse the whole of any foreign taxonomic system. They, too, pieced together those sources who accorded suitable taxonomic recognition to American subgroups. Like Thomas Jefferson, many academy naturalists envisioned a branched chain of being.[40] They argued that the taxonomic positions or "ranks" genera occupied upon these branches were ordered and that some day a true or natural taxonomic system would "place them all on equal footing."[41]

The work of Alexander Wilson (as well as that of his teacher William Bartram and Benjamin Smith Barton) had given American ornithologists the advantage of having a group of enthusiastic readers, willing to pay high prices for bird books. Subscriptions ranged from $180 for Charles Lucien Bonaparte's supplement to Wilson's volumes (originally $120) and $1,000 for Audubon's elephant folio volumes. Nevertheless, by 1825, Bonaparte was unwilling to organize his volumes according to a theoretical system. Instead, the sequence of the plates and text in his first volume reflected the order in which the western birds were collected and drawn by Titian Peale. If there was still room on the page, an extraneous small eastern bird, usually a female unrecorded by Wilson, was added by the engraver to the published picture.[42] Such haphazard presentation offered readers no systematic guide, and in a subsequent British edition, the work was reorganized by a Linnaean editor, Robert Jameson.[43]

Jameson's solution masked the need for taxonomic reform made evident in Bonaparte's original American edition. Furthermore, reviewers on both sides of the Atlantic considered Jameson's rearrangements to be a suppression of the American spirit.[44] Hailing the "impulse of genius" fostered by the American wilderness, they welcomed instead the study of ornithology freed from the "fog of technicalities thrown on it by the Linnaean school, and yet denser mystifications of ridiculous legends" by later "Cabinet Naturalists."[45] Apparently readers of American bird books were willing to waive the conveniences of systematic organization to enjoy the unfettered presentation of new studies. More careful attention to the text shows that Bonaparte had straddled the fence with his "intention to pursue a middle course." Like Say, he recognized the primacy of the "system of Nature," but he also justified the use of artificial distinctions:

If we proceed to the abolition of all artificial distinction between genera united by almost imperceptible gradations . . . the whole of these would be confused together; and, in fact, orders and classes would be considered genera, and even the vast groups, thus formed, would be still observed to unite inseparably at their extremes and we should finally be compelled to consider all living bodies both animal and plant, as belonging to one genus.[46]

Bonaparte chose a "middle course" because he lacked a reasonable alternative. He retained some artificial divisions but refused to fix genera in any arrangement which he felt had no basis in nature. This choice may have spared him critical censure among colleagues at the academy and abroad, but in the then "unsettled state of the science," it provided no solutions to problems raised by new western species and weakened the full impact of the discoveries he described.

Those botanical authors who collected in the West faced similar problems when they attempted large-scale publications. For example, in the preface to his first volume of 1818, Nuttall argued that many problems of American taxonomy were simply artifacts of the artificial or Linnaean method, which would disappear if plants were classified by "natural affinities." He also claimed:

The concussion of revolution whether in science or politics even to fulfill the most important object but little accord with our natural desire of harmony. . . . But we are at length inclined to believe, that the last and most perfect of systems, perfect because the uncontaminated gift of Nature, is about to be conferred upon and confirmed by the Botanical world. The great plan of natural affinities, sublime and extensive, eludes the arrogance of solitary individuals, and requires the concert of every Botanist and the exploration of every country.[47]

Quaker pacifism and Baconian disregard for individual genius are both evident in this perhaps the earliest American reference to scientific revolutions. In light of Maclure's later New Harmony project, the "natural desire of harmony" represents a curious choice of terms to discuss science. Nuttall clearly preferred a more natural system, but he organized his 834 genera around the sexual system of Linnaeus for "convenience and prevalence." Use of the artificial method forced Nuttall to make some awkward assignments, and as a consequence, his catalog suffered from the tensions of a double standard. The work received acclaim, but in 1818 it lacked the impact it might have had in the hands of a committed Linnaean.

Academy authors had hoped that the scientific foundation their work provided would withstand the "concussion of revolution," and they thought of their efforts as a lasting monument to the nation's biological wealth. In fact, Nuttall played no role in the taxonomic "revolution" he designated. Say died prematurely, leaving his *Entomology* unfinished, and Bonaparte returned to Europe never having made good his promise of "extreme reform." The "concussion of revolution" was certainly felt at the Academy of Natural Sciences, but the large publications of these authors demonstrated that the manner of presentation can inhibit theoretical development in a science. Because academy authors such as Say and Bonaparte could not make strong taxonomic statements, their texts relied heavily on illustrations for scientific credibility. Say admitted that an important function of his later illustrated work was to provide surrogate collections for readers who were isolated from the cabinets of scientific societies.[48] This purpose called for illustrations showing obvious taxonomic characters and little attention to background, shadowing, or animated posture. At the same time, the starkness of the page and the sole emphasis on the specimen's form gave undeserved realism to a method of classification that placed a premium on external structure and none whatsoever on internal anatomy or biological relationships.[49]

Academy naturalists were quick to recognize that "the chief defect of various magnificent European works" was "the unnatural stiffness faithfully copied from stuffed skins," although Wilson had also "Drawn from Nature" in that fashion.[50] Furthermore, participation in western expeditions did not lead to the development of naturalistic illustration. Despite the skilled *trompe l'oeil* and realistic still-life painting techniques perfected by other members of the Peale family, Titian Peale drew birds for Bonaparte's volumes which, as Audubon impolitely remarked, looked posed for portraits.[51] In their supplement to Wilson's volumes, Peale and Bonaparte were confronted with the task of continuity as well as novelty, but the undeniably handsome result of their collaboration improved neither classification nor illustration. To the contrary, expensive serial publications funded by subscriptions locked authors into an increasingly obsolete mode of taxonomy and did nothing to promote the "natural desire of harmony" among field naturalists and cabinet naturalists—those who traveled to collect and those who curated their collections.

The Denizens of Nature   5

Like the "more practical geologists," naturalists who traveled west attempted to confine their work "exclusively to the observation of facts."[1] After exposure to the social and educational ideas of the geologist Maclure, however, "fastidious avoidance of hypothesis" was not completely possible at the academy. Among field researchers in contact with Maclure only Rafinesque devised a way to classify the races of man in a comprehensive system in which American species were recognized as unique. His resolution was a nonhierarchical theory of physical evolution for all species, which were in his scheme distinctions of temporary duration, but Rafinesque unified his system of "permutations" in virtual isolation from the scientific community of which he had once been a part.

After the War of 1812, the disappearance of Indian groups in the New World posed a special problem, the understanding of which it was felt would yield scientific insight of large dimension. Like Bartram, musing upon the fate of the Indians following the American Revolution, later observers may well have wondered whether the maintenance of human species in the New World was a government responsibility. Academy naturalists could not answer this question unequivocally on the basis of their science alone. According to the rationale of Jefferson's arguments, the races of man

65

represented incipient species, but Philadelphia naturalists, many of whom were Quakers or admirers of the Society of Friends, avoided this conclusion in print. They did not seek to bolster what Bartram had termed "negligence" of the red race and the "woefull predicament" of black people.[2] On the Wyeth trek to Oregon, a private land venture with scientific staff funded jointly by the academy and the American Philosophical Society, John Kirk Townsend, the assistant naturalist, predicted that wasteful white hunters would soon decimate the bison herds "merely for the tongues, or for practice with the rifle." Townsend recognized the consequences of this "slaughter" for the economy of the Plains Indians dependent upon the herds, but the young doctor was even more dismayed by the effect of white contact upon the fishing villages of the Columbia River. "The depopulation here has been truly fearful," he reported. "A gentleman told me, that only four years ago, as he wandered near what had formerly been a thickly peopled village, he counted no less than sixteen dead, men and women, lying unburied and festering in the sun in front of their habitations. Within the houses all were sick; not one had escaped contagion."[3]

According to Buffon, the natives of North America were physically sterile, degenerate human types who brutalized their wives with slavish labor. Their negligence had permitted the continued deterioration of the climate and watery soil, which in large part accounted for the poorer quality of species engendered in the New World.[4] Although they rejected Buffon's thesis, naturalists struggled to understand the disappearance of Indian groups as a consequence of their place in the natural history of the New World. Townsend's traveling companion and senior, Thomas Nuttall, wanted to believe that nature functioned "with an omniscient benevolence." Compromising Jefferson's "benevolent persuasion" that extinction was not natural, Nuttall argued that the ability to learn enabled species to avoid extinction by keeping pace with environmental changes.[5] The inherent gradualism Nuttall discerned in nature guarded species, including the most intelligent North American creatures, the Indians, from the encroachment of European civilization. "Nature is not a cruel demon," he wrote in

1819, "nor delights in the accomplishment of destruction. Those who are fed by her frugal bounties are but seldom hurried into excess."[6] Late eighteenth-century Americans tried to explain the Indians' origins; nineteenth-century naturalists who went west tried to understand their extinction.[7] Although Townsend declined to speculate about the origins or taxonomic status of the Indians he saw, he was unequivocal about their bleak future: "The thoughtful observer cannot avoid perceiving that in a very few years the [Indian] race must, in the nature of things become extinct."[8]

Archaeological remains made interpretations of the Indians more confusing because there was no agreement that the ancient peoples who had built the earthworks found in the Southeast and Midwest were the same race as contemporary groups.[9] As a result, without serious scientific impediment, certain Indian mounds were said to contain the remains of antediluvian New World pygmies, a term indiscriminately applied to African tribesmen, orangutans, and chimpanzees, and the most sensible naturalists found themselves confronted with "pygmy" remains.[10] On June 12, 1819, Say and Peale left the Long Expedition party to examine the many Indian mounds or graves in the vicinity of St. Louis. They traveled to a local plantation, "where the Indian graves, said to be of a pygmy race of people are."[11] The graves they found were only three or four feet in length but contained bones of normal length which had been deliberately separated from the flesh. A day's digging across the river from the town of Lilliput satisfied them that "all the bones found here were those of men of the common size," and the two naturalists rejoined Long. Later the expedition's journalist, Edwin James, suggested that Lewis and Clark had unwittingly given the idea of Indian pygmies additional substantiation in their earlier retelling of a Sioux legend. In any event, Say continued to test the validity of comparisons between the Indian, the African, and the ape. He investigated the physical basis of the Indians' supposed want of ardor, their small size, only to find that "the Missouri Indian is symmetrical and active, and in stature, equal, if not somewhat superior, to the ordinary European standard," and he questioned their supposed brutish posture. He reported instead "perfectly upright carriage of the person, without anything of the swinging gait so universal with the white people," and one which he added "they imitate in their sports to excite the merriment of the spectators."[12]

During the long winter passed at the military cantonment near Council Bluffs, scurvy incapacitated more than 360 of the accompanying troops, but Say managed to assemble unique ethnographical data on the little-known tribes of the Upper Missouri.[13] Other notes were among his effects stolen by deserters.[14] Say's sections on the Indians were probably the most important contributions of the expedition's official report, but Say did not present any conclusions about the classification of the so-called red race. By the first quarter of the nineteenth century, natural history in the United States had arrived at the first of a series of paradoxes observation alone would not resolve. Western experiences like Say's or Peale's demonstrated that the natural history method of identification and collection could not determine the place of the American Indians within the context of organized zoology. Piqued by petty theft at camp, Nuttall grumbled that to assert all men were of the same origin "because they had all invented a somewhat similar clothing, is quite as futile as the same conclusion would be in consideration of their all being born naked."[15] Yet he never wrote the volume on western Indians that he promised readers of his travel book in 1819, and by 1823, a European judge traveling with Long's second expedition had classified the Sauk Indians with pygmies as a type somewhere between the Asiatic orangutan and the European white.[16] Skeptical readers had no convincing recourse in current natural history published by the best scientific explorers in the United States, and European authorities did not present a consensus of thought on the races of man.

Linnaeus had classified all humans as one species. Later, Johann Blumenbach (1752–1840) championed the theory of five world races based on skin color, which subsequent thinkers understood as the complex result of climate or the product of a particular set of environmental conditions. Jefferson had suggested that in America blacks might reasonably be expected to whiten over time.[17] He documented a case of a slave who had whitened with age, and

Charles Willson Peale painted a portrait of old James, actually the victim of vitiligo.[18] The notion of environmentally determined racial color was given additional support in 1799 when the nation's leading physician, Benjamin Rush, ascribed black skin to a form of congenital leprosy, an endemic disease he believed could be cured.[19]

These views were not shared by all. An advocate of the Asiatic origin of the American Indian, Samuel Mitchill believed in the unity of man and questioned the attribution of racial characteristics to particular biological environments. For him, the racial question was larger than the Indians, and he was willing to discuss his views in print. His charming *Picture of New York* is, like Jefferson's *Notes* on his favorite state, a guide to his own thought. Often called "the Congressional Dictionary," Mitchill was a political physician whose interests ranged from Antoine Lavoisier's ideas about animal respiration to the chemical causes of disease. Mitchill was an environmentalist in that he was an early advocate of chemical fertilizers. An enemy of urban filth, he urged stringent regulations for meat inspection "according to the usages of the Jews." In 1807, he recorded that in the city there were almost 72,000 whites and 4,008 blacks, of which half were free. Mitchill demonstrated that after 1756 the black population (then mostly slave) had declined by two-thirds.[20] For Mitchill, such a trend represented a fundamental desideratum of republican government. In another place, he suggested a new name for the United States, "Fredonia,"[21] which persists as town names in New York, Ohio, and Kansas and as the name of a present-day seed company. Mitchill called the citizens of Fredonia Fredes, and he related racial issues to the social environment created by human government rather than to the strictly biological environment. Departing from popular opinion, his thesis contained the germ of William Maclure's more radical claims for a new society.

Given the widely held belief that Indian languages held the answer to the whereabouts of their creation, Benjamin Smith Barton advocated the unity of man on the basis of linguistic studies. After one generation of esoteric philological pursuits, the number of races increased from five to as many as twenty-two.[22] On the

Wilkes Expedition, Charles Pickering (1805–78) took advantage of the squadron's route home to record the type locations for the human races he recognized from the South Sea islands to Egypt. Members of earlier expeditions also investigated the physical basis of racial distinctions. Both Titian Peale and John Kirk Townsend procured Indian skulls and other skeletal remains for collections curated at the academy by Samuel George Morton.[23] American interest in the subject of race mushroomed during the 1830s, and Morton delivered public lectures on topics ranging from the origin and diversity of the human species to the brain size of various races, racial abilities, the origin of black skin color, cranial characteristics of Negroes, and human racial hybrids.[24] By 1839, Morton associated cranial shape as determined by arithmetic parameters with racial type. The Quaker physician inadvertently opened the door for others to associate cranial shape with brain size and brain size with mental capacity and social station. Morton's magnificently illustrated *Crania Americana* was the last large-scale academy publication Maclure's generosity made possible during the aging patron's lifetime, but it is unlikely that Maclure saw this book, which, bearing his name on the dedication page, initiated an era of social dogma of the sort he personally deplored.

As the title *Crania Americana* implies, Morton's taxonomic model was Wilson's *American Ornithology*. Morton opposed the idea of Asiatic descent of the American Indians and, not surprisingly, argued against Lamarckian inheritance of environmentally induced characteristics.[25] His sequel, *Crania Aegyptiaca,* incorporated Pickering's notes and was dedicated to George Robbins Glidden, an English friend of Robert Owen. Glidden, like the southern herpetologist John Edwards Holbrook, admired Morton's quantitative approach and his anatomical emphasis on the skulls, which at 867 specimens ranked among the finest collections in the world.[26] Unfortunately, the publication of Morton's folio edition coincided with and was rapidly associated with a census of blacks and lunatics published by Dr. Joseph Clark Nott. Nott argued in public that the races were separate species and separately created, and in scientific circles, John Bachman (1790–1874), a southern clergyman and colleague of Audubon, singled out Morton's lec-

tures for attack on religious grounds.[27] Bachman vigorously defended the unity of man and, unlike Morton, lent this same energy
to his relentless defense of slavery.[28] Relying upon the single creation of man as described in the biblical book of Genesis, Bachman
attributed racial differences to environmental factors.

Morton, Nott, and Glidden believed in the great age of the earth
argued by many geologists. They were dissatisfied with the brev-
ity of the Bible's chronology, and in bold opposition to Cuvier
they asserted the existence of human fossils and predicted their
discovery before the end of the century. Morton insisted that since
man was the most domestic of animals, he like them produced
fertile crossbreeds that could become the progenitors of new
races.[29] In contrast to Nott's shabby polemics, Morton's initial
position was not inconsistent with either Quaker tenets or current
medical ethics. It was old-fashioned, however, and Morton's emphasis on cranial allometry can best be understood within the
scientific values of an earlier century, which placed numbers,
weight, and measure above qualitative concerns.

Morton's position on the human races evolved logically from
Jefferson's earlier suggestion that American species were unique and
distinct from those in Europe, the incentive for early studies of
plants and animals at the academy. Phrenology, Morton's new
science of man, renewed confidence in the overriding value of
quantification, but it also generated so many parameters of comparison that Pickering's *Races of Man* is unintelligible to even the
most patient reader.[30] Given the importance of illustrated publications at the academy, Audubon may well have influenced Bachman's
destructive attacks on Morton, the academy's corresponding secretary. Bachman was Audubon's collaborator, and his daughters married Audubon's sons. Out of loyalty to Wilson, the academy had not
supported Audubon's proposed bird book in 1824, and the charismatic "woodsman" was made an honorary member only after
Maclure's death in 1843. *Crania Americana* should have been the
climax of the publications Maclure financed, a culmination of ideas
first expressed in Jefferson's *Notes* and Peale's portrait gallery. Instead, Morton's work inaugurated a decade of racist debates to the
detriment of zoology.[31]

While naturalists with zoological interests argued over the human races, botanists organized around methodological issues which on the surface seemed less compelling. Although Rafinesque had warned as early as 1820 that without proper pasture management a valuable native cane grass would become extinct, the profundity of his observation paled before the drama of human extinction.[32] Townsend vividly described a not so "distant" time "when the little trinkets and toys" of the Indians "will be picked up by the curious, and valued as mementoes of a nation passed away forever from the face of the earth." He continued in this melancholy vein, "It seems as if the fiat of the Creator had gone forth, that these poor denizens of the forest and the stream should go hence, and be seen of men no more."[33] In comparison, Rafinesque's didactic remarks about the "passive life" of vegetables seemed to belong to a separate biology, which could "progress" without recourse to controversial topics. Adding that botany "has a great attraction for youth and for the female sex," Rafinesque, for once, shared a commonly held view: the hands of women "appear to be made on purpose, as it were, to handle these delicate objects, and to assort blossoms and flowers to adorn and set off their own blossoming charm." Rafinesque also executed a number of amorous drawings of his female students in Lexington.[34] His interest in the anatomy of their hands was clearly more than pedagogic, and other botanists recognized the efficacy and the economic rewards of teaching women.[35] While zoology became mired in the altercations of Harlan, Godman, Ord, Audubon, Morton, and Bachman, in botany the major obstacle to the introduction of new ideas was educational reform—changing the way in which women learned botany.[36]

The Natural Plan

6

In the twentieth century, it is almost incredible that early eigh-teenth-century readers of Newton's "method of fluxions," the still difficult calculus his physics required, never suspected that the plants around them were sexual. After Linnaeus, the proverbial lilies of the field had male and female parts that could be counted and classified, and his botanical system added to the Enlighten-ment world-view of numbers, weight, and measure a fourth com-ponent, sex. Since the Middle Ages, flowers had been associated in verse and art with feminine virtues, and despite their new-found sexuality, nineteenth-century educators continued to regard the study of plants—botany—as "peculiarly suited to the culture of the female mind" because it gave "delicacy to taste, brilliancy to fancy, feelings which are often lost amidst the rubbish of the se-verer sciences."[1] Apparently it did not occur to many of these teachers, mostly young male doctors who did not practice, that the fairer sex was perfectly capable of using a lexicon to translate botanical Latin into unambiguous English. In August of 1818, Amos Eaton warned the younger John Torrey of the wisdom of omitting certain technical terms from his lectures: "You speak of the rejection of the word Phallus as going too far, because few know its import. Miss Frances Hanshaw of Northampton, two of my Albany girls and three or four in Troy, are good Latin scholars and very cautiously search for the origin of the generic and specific names of all the plants they find. A single word of this kind would do much mischief."[2] Mischief aside, Eaton saw that the education of his "Trojans and Albanians" of upstate New York was essential to the "final success" of botany in the United States. In his eyes, the exposure of "the female part of our race" would lead to the addition of taxonomic botany to the standard curricula of schools and colleges, and by 1835 a less naive Torrey urged, "Let your female friends see that the Natural Method of Botany is not a big bear."[3]

Unfortunately, Eaton's commitment to female education blocked his willingness to change the plan of instruction he had, with the backing of Stephen Van Rensselaer, put into practice after 1824.[4] Furthermore, despite his useful goal of teaching science in relation to daily life, Eaton's stubborn refusal to acknowledge vulcanism in geology made all his accomplishments, including his famous *Manual of Botany* (1817), subject to extreme derision within the scientific community.[5] Nonetheless, his Rensselaer school produced a number of influential thinkers in the realm of education for women. Mary Lyon, the founder of Mount Holyoke Seminary, and Alice Johnson, the science educator, were both his students. Almira Lincoln, another protégé, praised Eaton as "the first public lecturer on botany" in the country, and just as Eaton had warned Torrey, her *Familiar Lectures on Botany* minutely explained for women the Latin and Greek roots of every technical term between its covers.[6]

One of the great intellectual contributions of eighteenth-century natural history had been the establishment of a common physiological basis for understanding plant and animal functions. Originally philosophical aspects of the Aristotelian hierarchy of the three human souls, the terms "vegetable" and "animal" gained a new identity in the comprehensive systems of Linnaeus, Buffon, and Lamarck. In his seemingly effortless prose, Buffon referred to animals as "awakened" plants.[7] At the same time, Erasmus Darwin, versifying upon the technical experiments of Albrecht von Haller and others, celebrated the supposedly animal quality of plants, and Darwin's work was admired by such American naturalists as Mitchill and Rafinesque, who were moved by a lesser muse to rhyme. In many ways, Lamarck's new word *biologie* represented the European synthesis of the plant and animal sciences, and William Lawrence (1783–1867), a Fellow of the Royal Society who introduced the term to English usage, was influenced by Erasmus Darwin.[8] Indeed, a copy of Darwin's *Phytologia* is one of the few books known with certainty to have been in the Bartram library, and given Benjamin Smith Barton's views on race, it is not surprising that this copy was presented by him to William in 1800.[9]

Colonial botanists had been hopeful entrepreneurs seeking new business opportunities for useful and ornamental species.[10] Even during the revolutionary period, the botanical activity of Philadelphia nurserymen existed primarily to meet British horticultural demands for new medicinal herbs or decorative plants.[11] As taxonomy became a means for identifying what was American as American, nineteenth-century naturalists took pride in the sheer numbers of new published species their work generated. They also recognized that, although the Linnaean method was convenient, it was limited and artificial. In 1803, Barton wrote that "the Sexual System of Linnaeus cannot be immortal," and he predicted its replacement with "a system more agreeable to the scheme of nature."[12] As early as 1758, Alexander Garden had written, "The man who gives the natural system must be a *second Adam,* seeing intuitively the essential differences of things," but he continued, "Where is this man to be found?"[13] Sixty years later, American naturalists would ask the same question.[14] Another problem compounded with time. European definitions of American species were not always useful because the characteristics given for the species were not sufficiently distinct from those given for the genus. American naturalists became more discerning. Jefferson had listed 109 American species in his *Notes on the State of Virginia.*[15] By 1791, Bartram had identified 215 birds in the Southeast, and Wilson described as new another 47 species. By 1813, the rewards of discriminating classification were a scientific reality for the naturalists of the early nineteenth century.[16]

In ornithology alone, combined discoveries claimed by Wilson, Ord, and Bonaparte (382 species) represented more than 6 percent of all the known birds of the world.[17] Bonaparte alone claimed 62 new American species (although only 2 were really new—Cooper's hawk and Say's phoebe).[18] Success was not limited to ornithology. Say's specialty was beetles, of which there are approximately 30,000 recorded species.[19] Say named at least 1,150 new species. In 1837, the best authority, Count August Dejean, listed 22,399 species of Coleoptera. If that number is taken as the world listing and if the same degree of error is assumed for both figures, Say contributed about 5 percent of all the beetles known three years after his death. Thaddeus William Harris, the sometime caretaker of Say's collections after 1834, published 994 species from Massachusetts alone. As for botany, Thomas Nuttall told John Torrey in 1820 that he had discovered approximately 300 new species of flowering plants in five years in the field.[20]

The voluminous nomenclature produced by these authors precipitated a crisis, for taxonomy could no longer be memorized as Linnaeus had recommended.[21] As old European common names were rapidly exhausted, resort to Greek and Latin increased to the point that many new American species had no vernacular names. The burgeoning terminology surrounding each branch of natural history necessitated Say's glossaries, Bonaparte's long addenda, Nuttall's textbooks, and Eaton's educational reforms.[22] Of course, this necessity for the expansion of language was not unique to natural history. Alexis de Tocqueville observed that in America "the continual restlessness of a democracy leads to endless change of language as of all else" and that "old ideas get lost or take new forms or are perhaps subdivided with an infinite variety of nuances."[23] Nuttall, Say, and later Titian Peale, however, were forced to defend their new names. A growing number of native-born American naturalists found this trend toward the proliferation of nomenclature alarming. They argued that "the general reader" was kept from "becoming as well versed in these matters as those who claim exclusive right to a technical acquaintance with the subject."[24]

After 1826 Rafinesque was the only naturalist in Philadelphia who attempted to explore the theoretical possibilities of classification based upon "an infinite variety" of taxonomic characters.

Like Torrey, Rafinesque realized that the systematics of North American plants needed reorganization, and he regarded the restoration of what he called the true or natural Linnaean method as his American mission. Taxonomy, he claimed, had been abused by "artificial classifiers" who "overwhelm us with useless names." Rafinesque also realized that even on its own terms, the Linnaean system could never be truly complete because the "essential character of plants can only become fixed when all Species of a Genus are known." Botanical progress required that "the whole earth is explored thoroughly."[25] The last edition of Linnaeus's *Systema naturae* had described 8,500 plants and 4,236 animals, and since only 193 of these were American, the challenge was obvious. By 1833, Rafinesque claimed to have collected 25,000 specimens representing 4,200 species and 5,000 varieties of North American plants, and before his death, he had issued more than nine hundred works on subjects ranging from plants to mollusks, Mayan ideographs, and the natural system of classification.[26]

Rafinesque viewed natural populations as a tool for the analysis of the antiquity of species and their historical relationships. According to his theory, populous species and genera would have to be both the oldest and the most variable. Rare species were of more recent origin. In the true system of classification, Rafinesque argued, related groups should be placed together. The first model he considered had been developed much earlier by Antoine Laurent de Jussieu. Genera were arranged in a triple series of columns—acotyledons, monocotyledons, and dicotyledons—with connecting bridges.[27] A second scheme, adapted from the British botanist John Lindley and followed by Torrey, involved triple, concentric circles which were divided into rays. The inner circle represented acotyledons; the outer circle, dicotyledons.[28] Rafinesque also rejected a third idea, a genealogical tree with three main branches. Two representations of the natural system he did favor involved a network of related species and a topological model following a "geographical plan" first suggested by Linnaeus and repeated by A. P. deCandolle.[29] Rafinesque claimed to have built such a model of the natural system at the Academy of Natural Sciences in 1815.[30]

By 1830 Rafinesque felt he had discovered a theory of speciation, which he outlined in a letter to Torrey: "My view of the
Subject after long reflexion is that Plants are Subject to *Natural and
Successive Deviations!* both Specific & Generic; but when perm,
effected these *Deviations* become to all scientific purposes N Sp &
G. which ought to be named."[31] By emphasizing naming and
nomenclature, Rafinesque was able to explain why New World
species necessarily differed from Old World counterparts.[32] In
1842, the same year that Charles Darwin penned his seminal
"Sketch," Americans did not regard Rafinesque's theory of "Successive Deviations" as the natural method of classification they
wanted. Because they had rejected the theory of degeneration,
many naturalists who modeled their careers on Wilson's rejected
any process of biological change.[33] In an effort to refute Buffon,
Wilson argued that species were permanent, and Jefferson would
not concede that any link of the great chain of being, once established, could be broken.[34]

Although none of Wilson's followers denied the possibility of
extinction, they could not supply a biological explanation that
avoided degeneration or self-destruction. These naturalists were
also unable to explain a related problem—geographical distribution. Nuttall, for example, was concerned with the distribution of
isolated species within a genus, but his rejection of separate creation forced him to take awkward scientific positions.[35] In the case
of the genus *Solanum,* he argued that since most of the species
(tomatoes) were American, the two European species (eggplants)
could not truly be natives of Europe.[36] He was wrong. Nuttall
thought, again incorrectly, that the spotty range of the catalpa tree
in the trans-Mississippi West could be explained only by artificial
propagation.[37] Rafinesque, considering the same problem, suggested that the Indians had brought the tree with them when they
migrated from Asia to America.[38] Isolation made extermination
understandable, and, like Jefferson, both Rafinesque and Nuttall
explained away difficult biological questions by resorting to
human intervention—the Indians.

The use of human intervention to explain isolated populations
climaxed in the work of Charles Pickering, whose studies of bio-

geography were admired by the usually critical Rafinesque.[39] In later life, Pickering devoted all his energy to the impact of man on the worldwide distribution of plant species.[40] Pickering's early work, just as unusual, was concerned with the restriction of species to characteristic regions, a problem that had first perplexed Nuttall in 1818. Pickering argued that each species "has been kept back and confined within narrow limits by causes," but the causes he recognized—water, soil, temperature, elevation, and geographical barriers—were environmental factors, not geological catastrophes as Rafinesque argued.[41] Like Rafinesque, Pickering imagined that the biological "laws" of distribution worked at two levels, those of the species and of the genus or family. Both men considered the taxonomic genus a real entity in nature, not simply an abstract tool for classification. Rafinesque went so far in this belief as to argue that permutation was a twofold process involving macromutations at the generic level and micromutations on the specific level.[42]

Concern with taxonomic explanations at the generic level impeded any true understanding of species in nature and shifted emphasis away from the definition of new species. By 1826, some zoologists were beginning to voice important objections to more systematic nomenclature. They wished to use their scientific organizations to "check the present mania for making new species often on slight foundations."[43] The interests of Say, Peale, LeSueur, and Nuttall no longer dominated the institutions they had done so much to promote, and by 1830 large publications were no longer a goal of academy members.[44] Exploring naturalists who were absent from Philadelphia had little incentive to return to the academy, and they were never again elected to the positions of leadership they had enjoyed there before 1826. Indeed, once eager field authors now expressed extreme caution about publishing. Say wrote to a friend in Boston requesting a certain gentleman "with your surname" to "be cautious, for the honour of our country, & be more sure before he publishes."[45] Say's concern was not premature. In his presidential address read before the New-York Lyceum of Natural History, James DeKay warned that "it is more meritous to extinguish a single nominal species than to establish a dozen new ones."[46] A zoologist, DeKay found support for his views among his peers. Godman—his friendships with Say and Peale notwithstanding—seriously questioned the scientific value of "imposing catalogues of NEW SPECIES, which at best might be little better than a string of barbarous new names applied to old and well known things."[47]

The "NEW SPECIES" referred to were more often than not variant populations that field naturalists designated as new groups. As a result, variation as such did not play a puzzling role in their science. On the contrary, the existence of variant populations within a larger, recognized division was, as Rafinesque reiterated, nature's reminder that naturalists must ever review established groups with an eye for subdivision.[48] Those who did not deem varying characters to be sufficient basis for new taxonomic divisions were forced early in their careers to confront the significance of variation.[49] For example, having described 332 species in the Linnaean fashion, Torrey, overwhelmed by the magnitude of the shortcomings of the artificial system, abandoned his undertaking after the first three parts. His failure must have been a source of personal embarrassment for the work was much anticipated by the press.[50] Consideration of variation, of course, brought with it a host of profound biological questions and a return to foreign authorities writing on American natural history subjects. Those authorities—William J. Hooker, Robert Brown, William Swainson, and Richard Owen—were British, and Benjamin Silliman, editor of the *American Journal of Science,* regretted this trend because cheap editions of English books undermined the market for American publications. Eager to maintain Wilson's tradition of quality, Silliman hoped for "a reciprocal recognition of the rights of literary property on both sides of the Atlantic."[51]

During the 1820s, Torrey, for one, did not push on to perfect a system that would be rendered obsolete by the natural system, nor did other botanists with academic interests seek to build an American science on a foundation of comprehensive books. Instead, their focus was regional, and they were eager to amass local collections or specimens that could be stored until the proper methods for classification became available to American naturalists.[52] Repeating earlier criticisms of Buffon, their opponents accused naturalists who were not professional western explorers of working in the closet. These closet naturalists were American-born practitioners whose range of collecting expertise was confined to the American Northeast, and at this stage in their careers, their attitudes were not infrequently compromised by provincialism. Torrey wrote to Gray that he "would rather have a new Yankee grass than a new palm from any of the Mexican states."[53] The newness of their taxonomic materials and the novelty of their publications were not of primary importance. Torrey and his correspondents were often content to work with well-known materials or to arrange materials that others sent them in contrast to those naturalists who worked on the belief that it "is better to have two names for an object than no name at all." In their absence, western naturalists began to feel that their original designations were no longer being honored, and the inevitable nomenclatural disputes that accompanied discovery quickly moved from a scientific level to the realm of professional and even personal ethics.[54]

Curtailing published nomenclature did not lead to new insights or the organization of American natural history initiated in 1831 by the first American edition of John Lindley's *Introduction to the Natural System of Botany* supplemented by Torrey.[55] Torrey's edition of Lindley's natural method was immediately praised by reviewers in the scientific press with one exception—field naturalists.[56] These men expressed objections to the changed status of field work which they felt adoption of Lindley's system entailed. Even Torrey's old friend Eaton denounced foreign reforms that were not the "progressive outcome of researches" in the American field.[57] Rafinesque disliked Lindley's "anatomical" or "closet" approach, which relied upon indoor microscopical inspection, not gross or unaided outdoor observation.[58] Nuttall complained that field researchers were being treated like underpaid collectors by former colleagues.[59]

Of these critics, only Rafinesque had a real theoretical need to emphasize the importance of field study for the observation of permutations. Since the constant origin of new species through permutation permitted Rafinesque to dispense with independent creations and to reaffirm the validity of American forms under distinct new names, he believed he had fulfilled the goal of a generation of naturalists working in this country.[60]

Unfortunately for the self-styled "Galileo," Gray shared this same vanity.[61] In addition to Rafinesque's obscure natural method, there were many others to choose from that did not involve evolution, and by the end of 1835, Gray was under the sway of DeCandolle. In November of that year, the old schoolteacher Eaton visited Gray to assure the younger man that he wished "to keep pace with *published discoveries*." Eaton claimed that Gray replied, "Take thy shoes from off thy feet, and put on the sandals of the *Natural Method*, and *we* will begin to hear you."[62] This "small specimen of his urbanity," as Eaton phrased it, ensured his "contempt" for Gray, but the tide of scientific opinion had already changed. The sandals of the natural method were not in Nuttall's words "moccasins soaked" with the honey of wildflowers.[63] Despite the considerable contributions and popularity of Eaton, Rafinesque, and Nuttall, one by one, Gray undid their scientific reputations through either personal confrontation or harshly worded published reviews. Even mild-mannered Torrey dismissed Eaton's students as "mere smatterers—not even tolerable collectors."[64]

Needless to say, Torrey's natural method did not become the permanent method for botany. Lindley's system of circles is all but forgotten, but Torrey's step was a bold one for it separated botanists who were strictly herbarium keepers from those who were primarily educators or researchers with zoological interests as well. In truth, Torrey's endorsement of a foreign system foreshadowed his student Gray's Darwinism, and as one consequence of Torrey's action, Gray traveled abroad to examine American type specimens housed in foreign collections. In England, he met Charles Darwin and Joseph Dalton Hooker, and it was his subsequent correspondence with the botanist Hooker that permitted his role as Darwin's American ally.[65] Gray later wrote that Torrey foresaw that the natural method was destined "to change the character of botanical instruction."[66]

Torrey opposed innovations in the interest of stability, and it was not until his third paper on the plants collected by Edwin James on the Long Expedition that he used a natural arrangement.[67] By 1828, after study of materials collected on the Long Expedition, Bonaparte suggested that Nicholas A. Vigors's natural system be adopted for the classification of birds.[68] Bonaparte made the same transition as Torrey's student Gray from the natural system to evolution, but sooner, and by 1850 he was openly an evolutionist.[69] With the exception of Rafinesque, in 1826 or 1828, natural systems did not imply evolutionary change in the minds of advocates. Natural systems did imply professional and instructional changes. Profound taxonomic reform would, as Eaton argued, entail an "awful catastrophe," an educational revolution with new teachers, new teaching methods, and new textbooks. Maclure was ready to undertake such a revolution for he saw natural history as a force for social reform.

By 1828 William Lawrence had applied agricultural concepts to the natural history of man, and Maclure may have heard Lawrence's lectures in Great Britain before they were rapidly disseminated in the United States. Lawrence, later the well-traveled surgeon to Queen Victoria, described human races as varieties of the same species and controversially attributed noticeable physical differences such as hair type to the idea that like sheep some human varieties were wild and some were domestic. Slight differences in nature became accentuated varietal types through "instinctive aversion," his concept of sexual selection. Lawrence also emphasized the role of sterile hybrids in speciation and the "allotment" of species to certain regions of the world. In this way, he brought together Buffon's notion of sterile hybrids with Linnaean geographical distribution based upon the idea of creation as a garden. Lawrence's agricultural explanations for the process of race also revitalized the concept of degeneration: because "man is eminently domestic," he exposes himself "more than any other animal to the causes of degeneration."[70] Maclure recognized principles of degeneration at work in human populations, but the causes he identified by 1819 were ignorance and poverty, not biology. Maclure shared the view, later given grander articulation by Morton, that man was the subject of his own domestication. In Maclure's opinion, people were more often than not victims of their forms of government. Like Jefferson on the subject of skin color, Maclure hoped a controlled educational environment could reverse social and economic degeneration, and he attempted this experiment at the schools of New Harmony, Indiana.

PART THREE

New Harmony

A New Organon

In the area of early nineteenth-century American natural history, Francis Bacon's emphasis on collection, group effort, and written "histories" rather than mathematics met with success. The American endorsement of the Baconian approach to science has received much scholarly attention, although historians have tended to neglect the influence of another aspect of Bacon's thought—his scientific utopianism. Bacon wrote his famous *New Atlantis* late in life, and the work was published posthumously in 1627. As is well known, the *New Atlantis* describes Bensalem, an ideal commonwealth discovered by a group of voyagers en route from Peru to Japan. The inhabitants are a benevolent, scholarly people who pursue useful research in well-equipped laboratories in towers half a mile high and enjoy artificial lakes, zoological parks, and other improvements.

During the early seventeenth century, Bacon's utopianism was paralleled in part by the mystical Rosicrucian movement, which called for a new learning couched in a utopian format and demanded Paracelsian reforms in science, education, and medicine.[1] Rosicrucian thought made some inroads into the real new Atlantis, the New World, in 1704 with the arrival of Christopher Witt, a physician, musician, botanist, and mystic. Dr. Witt was the neighbor of James Logan, William Penn's secretary, who became mayor of Philadelphia and then chief justice and acting governor of Pennsylvania, and Witt's remarkable library and his interest in plants attracted other Quakers as well, including John Bartram, who politely endured Witt's conversations on magic in order to read his books.[2]

The profound influence of Witt's ideas upon his contemporaries has not been studied, but elements of his ideas persisted through the eighteenth century, and pseudo-scientific mysticism associated with the original Rosicrucian movement resurfaced in the nineteenth century in American projects. The Indiana Harmonie project of George Rapp (1757–1847), for example, acquired a mysterious stone slab imprinted with ancient human footsteps, and the spiritualism of Owen family members certainly affected the later New Harmony project at the same site. The small, cooperative societies described by Bacon and the educational reforms urged by the Rosicrucians were effected in varying degrees at the Moravian schools and academies run by such botanists as Lewis de Schweinitz and Amos Eaton and by the Quaker and Shaker schools, which here and abroad shaped Maclure's educational goals. Indeed, much of the original furniture at New Harmony was of Shaker origin.[3] Maclure anticipated science prospering among the "Industrious Producers" of the frontier, and the participation of notable members of the Academy of Natural Sciences (latter-day counterparts, it might be said, of the sixteenth-century wayfarers to Bensalem) embodied the epitomic refutation of Buffon's theory of degeneration in the New World. The dissemination of scientific knowledge, Maclure believed, would lead to the material improvement of every aspect of American life and eliminate the mental degeneration Maclure associated with slavery, poverty, and most organized religion. Unlike Bacon or other utopian writers, Maclure was able to implement these ideas during his lifetime, and he was willing to finance this ultimate test of the "science of society" on a scale that reflected his large, self-earned fortune.

The association of science through education with the goals of a good government or, in Clinton's case, with the practical politics of a new nation, was not difficult.[4] Thomas Jefferson was able to organize the University of Virginia among a people who enjoyed the second highest level of literacy in the world, surpassed only by that of Prussia. Greek and Latin, the basis of a traditional education in the classics, were no impediment to science. To the contrary, the advancement of natural history required a good grasp of these scholarly languages because the discipline relied on Greek and Latin nomenclature. Despite the compatibility of science and the classics, however, natural history was not well defined in college curricula. In 1799, Benjamin Smith Barton found it was necessary to defend natural history instruction in the colleges so as to expose common observers to its fundamentals and to bring these before the "ingenuity of philosophers."[5] He urged "innocent and useful zeal," but his promotion of American studies differed from that of the next generation in at least two ways. First, although Barton collected and published in this country, he also published abroad.[6] Second, because Barton realized that "it is in [man's] power to increase or diminish the number of animals and vegetables about him; and even to destroy *whole species*," he urged cautious pursuit of natural history for practical or political ends. Without an academic or theoretical basis, the tools of natural history, he recognized, could be misused "in a system so complex, and so difficult as this."[7]

By 1800 the number of colleges in this country rose from nine to twenty-six, and by 1828 fifty were in operation.[8] Most of these newer colleges, located in the Northeast, began as small, isolated Protestant schools in which the classical tradition was partner to rather than servant of theology. Although there was no organized antiscientific feeling at these colleges, they were not able to create the spirit of scientific deliberation Barton desired.[9] After the War of 1812, the proliferation of academies offering scientific education at the secondary school level graduated better-prepared students, but the colleges still presented such similar curricula that they were in some ways indistinguishable. With the notable exception of Maclure, few educators equated scientific progress with technological advances or national defense, and science and classical knowledge remained fused within the liberal arts.

Having applied the Wernerian system of mineralogical classification to the soils of his new country in 1808 Maclure remained appalled that most people went through life entirely ignorant of the stones upon which they trod. Twenty-five years later, his sentiments were given bizarre expression in a painting by Thomas Cole that shows a huge goblet towering ominously over a landscape across which the inhabitants move, oblivious to its significance or even its presence.[10] So gradual are the effects of the overflowing goblet that they cannot be observed by those whose lives they shape. Maclure's survey of the national mineral wealth may have been a timely reaction to Jefferson's embargo, but his efforts—his published letters, the revision of his survey in 1817, and his support of amateur societies—continued to provoke public interest in mineralogy and its important implications for the development of commerce and defense. By contrast, Maclure's intellectual voyage and the motives behind his transition from business to geology and from geology to social reform remain poorly understood and in a sense are as challenging today as they were to his puzzled contemporaries.[11]

In the closing years of the eighteenth century, Condillac's philosophy was taken up with special eagerness by the French *ideologues,* the most radical of whom wished to apply his ideas about language and communication to educational programs designed to secure the advancement of knowledge and the perfectibility of man. Maclure was exposed to the thought of the *ideologues* in France, and later one of the teachers he hired translated the *Logic of Condillac* to illuminate the plan of education already in effect in Maclure's Philadelphia schools. During the 1830s, Maclure reviewed his objectives for the academy's corresponding secretary, Samuel G. Morton. Maclure sought a moral revolution through the equalization of power. He desired social improvements through modern science and hoped to implement useful rather than erudite publications. These publications were to be inexpensive and, eventually, self-sufficient. Finally, he hoped for a freedom from fear, which Maclure labeled "that cancer of the mind."[12]

Although little is known about the development of Maclure's social ideas following his retirement from the mercantile firm of Miller, Hart, and Company, some of his goals coincided with commonly held expectations. For example, an elementary botanical lexicon of 1820 claimed that "an intimacy with natural objects" serves "to produce an accuracy of discrimination, which is the foundation of correct taste, the essence of sound judgment, and the object of every judicious system of education."[13] The real issue was just how far this "intimacy" with nature should be carried. Eaton's student Almira Lincoln valued natural history for the instruction of young people, but she questioned the wisdom of natural history as an adult profession. She noted that "naturalists to the great discredit of the science have sometimes shown an unhappy tendency to skepticism."[14] That skepticism, often a euphemism for atheism and materialism, had given rise in Philadelphia both to the Academy of Natural Sciences in 1812 and to Maclure's education projects. Although Maclure's stated philosophy became far more radical than that of his associates, after his election as president in 1817, his patronage of the academy was immediate and long-lasting.

From the very first, however, Maclure's activities at the academy were not purely scientific. During the winter of 1817–18, he sponsored an exploring party to East Florida, which included the zoologists George Ord and Thomas Say, as well as seventeen-year-old Titian Ramsay Peale. Although all four men, even young Titian, were officers of the newly incorporated Academy of Natural Sciences, the exact expectations of the academy are unclear.[15] Other than Say's remarks to scientific colleagues and Titian's few letters home, no contemporary narrative account of the trip survives, and Peale's description dictated as an old man gives no explanation of the explorers' purposes.[16] Maclure's survey of the region east of the Mississippi River had already mapped their destination, the mouth of the St. Johns River, and Andrew Ellicott's earlier survey of the St. Marys River area nearby had been published in Philadelphia by the firm of Budd and Bartram.[17] The group also seemed initially oblivious to the potential danger of the undertaking. Although the Treaty of Paris had returned East and West Florida to Spain in 1783, Spanish influence was weak, and Spain was unable to protect the eastern waterways from renegade slaves, marauding Indians, or pirates.[18] Consequently, the United States government received many protests of disorder on the Florida frontier, and on December 23, 1817, two days before the naturalists departed from Philadelphia, two hundred troops landed on Amelia Island at the mouth of the St. Johns River to hold it until the anticipated United States annexation of the Floridas.

Maclure had invited Say at the last minute to follow in "the track of Bartram," but the Floridas possessed too little mineral wealth or topographical relief to justify a geological survey so hastily arranged and so hazardous. As the Philadelphia naturalists were to discover, American military presence did not allay Indian hostilities on the peninsula or safeguard coastal travel, but the party of eight, which included three hired sailors, Maclure's servant, and Ord's hunting dogs, visited the Georgia Sea Island cotton planters (the source of Owen's wealth) and made its way up the St. Johns River. Say wrote naively, "We entertain no fears from hostility of the Indians, we could even repel the attack of a few of them."[19]

Concern for his own person or for the safety of the men with him compelled Maclure to remain at Fort Picolata on the upper St. Johns River while Peale, Ord, and Say made their way east along the swampy King's Road to verify their royal passports at St. Augustine. Say's correspondence indicates that after learning that Indian raids made further venture up the St. Johns River imprudent, they backtracked with the hope of exploring the mouth of the Mosquito River, the site of William Bartram's ill-fated plantation. To their dismay, they were again thwarted by reports of dangerous Indians, and once again they decided to pursue the upper reaches of the St. Johns River, which Bartram had explored so profitably more than forty years earlier. After meeting a plantation owner whose son had just been murdered by angry Indians, they stopped briefly for Say to gather land snails in the ruins of the old Fort Picolata and then "departed from that place in good time." Ord, an affluent ship chandler, left the party at St. Marys and returned to his business in Philadelphia. The other members elected to sail north, collecting along the Sea Islands to Savannah, where they disposed of their sloop and sailed to Charleston in the *Rambler.* Maclure, who had suffered violent seasickness, did not

sail but returned by steamboat, and the little scientific party recon-
vened in Philadelphia at the end of April 1818. Peale brought home
some birds for his father's museum, Maclure's geological findings
were negligible, and Say, who managed to collect at least nine new
species of shells which he described the next year, concluded with
pacifistic sentiments characteristic of his Quaker background:
"Thus, in consequence of this most cruel and inhumane war that
our government is unrighteously and unconstitutionally waging
against these poor wretches whom we call savages, our voyage of
discovery was rendered abortive."[20]

One well might wonder why the Philadelphia naturalists chose
to explore Florida at this time. The celerity with which Maclure
was able to receive a Spanish passport for the trip surprised Say,
who commented that the trip was planned on extremely short
notice. The geology did not merit this dangerous trip, for the
region is so young that it lacks the variety of minerals found in
older regions of North America with metamorphic and igneous
exposures. Furthermore, it is unlikely that the mineralogist Maclure was aware of the vast phosphate reserves to the south that
furnish the state's modern industry. Did Maclure's haste stem from
his reading of the *Philadelphia Aurora and General Advertiser* for
December 8, which described the provisional government of Vi-
cente Pazos and other supporters of the Spanish-American "pa-
triots" on Amelia Island in the mouth of the St. Johns River?
Maclure was committed to the cause of Spanish freedom in the
New World, and shortly afterward he purchased ten thousand
acres of expropriated church lands in Alicante, Spain. He supported the new liberal constitution to the tune of $60,000 sunk into
an experimental agricultural and industrial school that never
opened before the royalists regained power.[21] Thus by the time
Owen advertised in this country for backers for New Harmony,
foreign observation and large cash investments were not new elements in Maclure's life.

In 1805, he had visited the school of Johann Heinrich Pestalozzi
(1746–1827) in Yverdon, Switzerland, which for five years attracted much press attention in the United States. Endorsing the
philosophy of Condillac, Pestalozzi believed that clear thinking

was dependent upon accurate observation of actual objects, and his methods were an attempt to put into practice the educational theories of Rousseau.[22] Pestalozzi emphasized such student activities as drawing, model-making, collecting, and field trips—all activities which Maclure recognized would lend themselves well to the progress of science. "To collect facts," he later wrote, "without being warped by an attachment to system, is the surest mode of advancing geology, as well as all other sciences."[23]

Maclure was eager to popularize the Pestalozzian method in the United States, and toward this end Pestalozzi recommended an instructor from his school, Joseph Neef, a veteran of the Napoleonic wars. The Pestalozzian system received good press in New York State, where aspects of it had been funded by the legislature, but in Pennsylvania, Neef's atheism resulted in the closing of Maclure's school in Delaware County in 1814. Neef subsequently moved his family to a farm in Kentucky, until Maclure relocated them in the New Harmony school system. By 1824, Maclure was planning a school for the community children from two to five years of age to free them from parental influence. His teachers were to employ "prints, instruments, representations, books" to facilitate "the giving of distinct and accurate ideas to children, at a much earlier period than had yet been practiced."[24] He hoped "by imprinting the image on their minds directly from the object itself" to give "a true and exact representation of it, in place of the old, irksome, fatiguing and imperfect way of description." Maclure's means, if not his final goals, corresponded exactly with those advocated by proponents of the natural method of classification.

By 1818, Nuttall, Rafinesque, and Say had addressed the need for a natural method of classification. By 1825, they were joined by Torrey and Bonaparte, but these very different naturalists all hesitated to take action for the same reasons. They realized that there was no audience for the natural system, no group of students trained to put it into effect. Maclure's system of education offered a way for Say to create this audience. Furthermore, despite the rough conditions of the frontier, the site was a good one for natural history, especially for geology and river productions—fish and bivalves.

The odd spoonbill sturgeon or paddlefish (*Polydon spathula*) was not infrequently found, and the now extinct Carolina paroquet (*Conuropsis carolinensis*) spent the winter there living on the fruit of the sycamore trees. Maclure needed to offer little additional incentive to staff his schools with the best scientific teachers of his day. At the University of Pennsylvania, for example, the faculty of the undergraduate Department of Natural Science—Thomas Say, William Keating, and Barton's nephew W. P. C. Barton—were so poorly paid and so underused by 1821 that both Say and Keating traveled west with Long's second expedition.

Low salaries were not unique to Philadelphia, and Maclure's patronage, whether at the academy or on the frontier, had an impact upon other naturalists active from 1817 to 1840, even those who did not agree with his social theories. After twelve years "vegetating at Harvard," Thomas Nuttall gladly relinquished his charge of the botanic garden in Cambridge to resume a life of exploration under the auspices of the academy and the American Philosophical Society.[25] In addition to expedition appointments, the few salaried positions offered by scientific societies were among the highest paid jobs in natural history in the United States, and during the 1830s, the entomologist Thaddeus William Harris hoped to leave his post at Harvard College to assume the librarianship of the academy, a job Pickering received instead. Besides his ties to organized scientific societies, Maclure maintained at least three experimental schools in the Philadelphia area, and he had the finances to open others. In 1825, many of the academy members and scientists in other cities were known to him personally and respected his leadership. With the possible exception of his Mexican endeavors, all Maclure attempted at New Harmony could have been implemented in Philadelphia and perhaps with greater success, but Maclure identified his educational goals with social unrest. He sought to establish his projects in areas of potential political change, and his costly interests in Spanish schools at a time of constitutional revolution, his death in Mexico, and his bequest of $80,000 worth of books to the Indians of Indiana are all aspects of his commitment to reform.

Unlike other American educational movements during the

1820s and 1830s, Maclure's New Harmony schools were coeducational. While his botanical contemporaries debated the efficacy of teaching the sexual system of Linnaeus in female seminaries, young women and men worked together on the scientific press Maclure subsidized at New Harmony and benefited from the library of more than two thousand scientific volumes he had purchased in Europe. Until more of the facts of Maclure's personal life are known, the details of many of his educational projects will remain obscure. His interest in Owen's project in New Lanark and Pestalozzi's Yverdon schools first became manifest during the period he served President Thomas Jefferson in 1803 as a liaison for American citizens whose foreign properties had been confiscated during the French Revolution, but we can only speculate about Maclure's motives in Spanish East Florida. Did he hope to start a frontier school among the displaced Indians, runaway slaves, and South American rebels in what was then the planned—but unrealized—bluff town of St. Johns? A few years later, was he influenced by Frances Wright? Having established a community at Nashoba, Tennessee, to recondition freed slaves for return to Africa, Wright went to New Harmony, where she made a considerable impression on her new colleagues. One of the community's early members of the Council for Domestic Economy traveled to Nashoba with Wright and married her sister.[26] It is possible that Maclure, using Wright's ideas as he had used Owen's, went to Mexico to restore Mission Indians to their aboriginal status as noble savages. Although he had ceased to be an active field researcher by 1825, Maclure had not retired from the world of ideas, and given his mineralogical orientation to geology, the old Wernerian's activities in Mexico merit scholarly scrutiny, for the war that followed his death between Mexico and the United States centered upon mineral rights.

A great enemy of ignorance, Maclure was a man of sanguine generosity. Indeed, Rafinesque's description of himself "like *Bacon* and *Galileo* somewhat ahead of the age and my neighbors" far better suited Maclure. Certainly neither social station nor advancing years held Maclure back from the challenge of an intriguing experiment, and he became a martyr to his new "science of society." In March 1840, having decided to return to New Harmony, he collapsed en route and died in the Mexican village of San Angelo, a fortuitous association the benevolent materialist would not have recognized.

The Vortex of Experiment

8

Harmonie, a community of thirty thousand acres on the Wabash River in southwestern Indiana, had been founded in 1814 by a German Pietist named Johann George Rapp. Seeking religious freedom to prepare for the millennium, Rapp successfully petitioned President Jefferson to solicit land in the "Western County," and on Independence Day of 1804, he and three hundred followers arrived in Baltimore. Within a decade they had transplanted orchards from an earlier settlement in Butler County, Pennsylvania, and set to work planting fields and vineyards.[1] The industries they began were successful and were soon servicing the frontier with goods bearing the trademark of the Rappite Rose.

Most of the Rappites professed celibacy, although this stricture does not seem to have been enforced. In Germany they had been celibate for purposes of economy. In 1798, the English economist Thomas Malthus had grimly concluded that poverty was unavoidable because populations increased by a geometrical ratio while the means of subsistence increased by an arithmetic ratio. By 1803, the year the Rappites left Württemberg, Malthus suggested the efficacy of moral restraint, celibacy, and late marriage on population growth in his revised *Essay on the Principle of Population,* a work both Charles Darwin and Alfred Wallace named as decisive in their independent formulations of the theory of natural selection.[2] Malthus's work stimulated other readers as well. The poet George Byron described Rapp as a Malthusianist in his epic *Don Juan* (canto 15, stanzas 35–38), and Owen, greatly impressed by the material prosperity of celibate communities like the Shakers, described their way of life in his 1813 *New View of Society*.[3]

Outsiders often interpreted the success of Harmonie in material terms rather than along the spiritual lines of its founders, but without the enthusiastic accounts of English travelers Owen probably would not have bought the village. In contrast, the Rappites' countryman Paul Wilhelm, duke of Württemberg, was critical of their "misconceived desire for improvement or a false conception of liberty."[4] In 1824, the Rappites left New Harmony to return to Pennsylvania. They founded the town of Economy outside Pittsburgh and in 1819 sent Richard Flower to England with a commission to sell the Indiana village.[5] Robert Owen first wrote to Rapp from New Lanark in August 1820 asking about the "rise, progress and present state of Harmony."[6] With a Shaker community nearby at Vincennes, ten thousand acres under cultivation, nineteen detached farms, a village center, factories, and machinery, Rapp's asking price of $150,000 seemed a bargain compared to $480,000, the estimated cost of starting a similar community in England.[7] Owen apparently was unaware that the American press had already attacked his views on religion.

Of humble origins, Owen was the first manufacturer to import in 1791 what François A. Michaux described as "the best of cotton," known in the French trade as "Georgia cotton" and in England "by the name of Sea Island cotton."[8] Owen induced his partners in the Chorlton Twist Company, Barrowdale and Atkinson, to buy the mills of New Lanark from David Dale (1739–1806) for £60,000 and shortly thereafter, he married Dale's daughter Caroline.[9] Owen attempted to reorganize the mills. He hoped to improve housing, sanitation, and educational facilities for the thousand persons connected with the mills, and his "Infant school" became the first nursery of its kind in Great Britain. His efforts met with success. Even throughout the United States embargo on cotton in 1806, Owen was able to pay his workers full wages.

Despite New Lanark's financial profits, Owen's partners grumbled about his methods and offered the mills for public sale in 1813. Owen bought them back for £777,000 with six London partners, including the Quakers William Allen and Jeremy Bentham. Owen's new firm brought together the same attitudes of British reform and the Society of Friends that gave rise to the Academy of Natural Sciences in Philadelphia. Indeed, Bentham's nephew George Bentham became a botanist of repute and worked with Asa Gray in 1850 on the plant collections of the Wilkes Expedition. [10]

A New View of Society was Owen's first and most important work stating the principles behind the educational and social reforms of New Lanark. [11] Owen claimed that men and women should be placed under the "proper" influences in their earliest years, and he argued that human character was formed by circumstances over which people have no control. Owen also insisted that human beings were neither the guilty nor the redeemed creatures organized religion purported them to be. In his *New Moral World,* Owen claimed that "human nature is a compound of animal propensities, intellectual faculties and moral qualities," which are mixed differently in each individual by a "power unknown." Nature's laws required that physical, mental, and moral feelings should be exercised with temperance, which was possible only if all inferior circumstances were removed from the environment. Happiness was thus defined for an individual by the harmony of physical, mental, and moral proportions in the proper environment. To achieve the latter, "the science of society" required knowledge of laws of human nature, abundance and comfort without waste, and the proper distribution of wealth. Owen's last law, the practice of rational religion, became a thorny issue even with his new partners, and after Owen left for America, the curriculum of the schools of New Lanark was so changed that upon his return to Britain in 1828 Owen severed his ties. [12]

In 1817, Owen announced that the basic cause of human distress was mechanization, the competition of human labor with machines. The only effective remedy was the united effort of men and the subordination of machinery. Owen suggested that parcels of land from one hundred to fifteen hundred acres be settled by twelve hundred persons. Each family would have a private apartment in a large communal building, but the community would be responsible for all children over three years of age. The culmination of his ideas, which were receiving a wide audience, was the purchase of Harmonie. In the meantime, thousands of tourists visited New Lanark, and on July 30, 1824, shortly before Owen left the model village for the United States, their numbers included William Maclure.

By November 1824, Maclure was shocked by the "plots and conspiracies" of the privileged against the productive classes in Spain and alarmed by the extent to which scientific publications discouraged common readership; he advocated publication of easy-to-read extracts printed on inexpensive paper made of "straw."[13] Maclure, by now a Malthusian of sorts, was "thoroughly persuaded that the geometrical progression of improvement and civilization" would support the public dissemination of scientific ideas.[14] Sharing Owen's opinion that "man is a *bundle* of habits—the child of surrounding circumstances," he wrote to Benjamin Silliman that education distinguishes man from "the brutes."[15] At New Lanark he was especially impressed by Owen's ability to reform adults after they had been filled with prejudices, and he was obviously inspired by the great success "Mr. O has obtained in the only fair and impartial experiment ever made on the great mass of industrious producers of material wealth." In another published letter to editor Silliman, Maclure declared that there were only two things in Britain worth imitating in America—infant schools and mechanical institutions.[16] Again he noted the cheap publications of the latter in editions of one hundred thousand at a time, but although he announced his intention to return to the United States in the spring of 1825 and briefly described Owen's purchase of Harmonie, he did not indicate to Silliman that he expected to participate in Owen's New Harmony project.

As a rule Silliman's New Haven journal did not discuss political or social matters, and it is thus noteworthy that he published extracts from Maclure's letters. Silliman, a geologist, may have acted out of deference to the benefactor of his favorite science, geology, or because Maclure was a well-known public figure, Silliman may have thought Maclure's exposition of Owen's ideas would interest his readers. Owen was already actively recruiting followers in other American cities, including Philadelphia and Washington, D.C., and on February 25 and again on March 7, 1825, he delivered addresses before the president, the justices of the Supreme Court, and both houses of Congress. The ideal model for New Harmony he proposed was a quadrangle, one thousand feet square, enclosed by an elaborate gothic construction, which Robert Dale Owen later defended as "organic" architecture.[17]

Sometime during 1825, Maclure decided to make New Harmony the center of educational reforms already in effect at his schools in Europe and Philadelphia. Limiting his liability to $10,000, Maclure agreed to finance the schools and to supply teachers, scientific equipment, and the library, much of which arrived at New Harmony aboard Maclure's *Philanthropist*.[18] The "boat-load of knowledge" also brought from the academy Say, LeSueur, Troost, Say's close friends John and Elizabeth Speakman, and the artist and engraver John Chappelsmith.[19] Having been impressed by James Fenimore Cooper's descriptions of the American West (which owed much to Say's and Peale's contributions to the published first Long Expedition report), Owen's son Robert Dale Owen (1810–77) also joined the community at this time. This group helped to draft the society's constitution, which claimed as self-evident the desire for happiness, and the community's newspaper, the *New Harmony Gazette*, declared 1826 to be "The First Year of Mental Independence."[20]

Owen's venture failed, and within a few weeks New Harmony was filled with dissension. Many who had joined were freeloaders, and Owen's latitudinarianism became increasingly unpopular among the sincere Christian members. In March the "boat-load of knowledge" resigned from the main village to form a splinter group, but, given uncleared lands to settle, they soon returned. Three months later, the same group seceded with nine hundred acres and formed the Education Society with Maclure as their patron. Their dissatisfaction was not unique. Other dissatisfied

Harmonites organized Macluria and Feiba-Peveli (a name derived from the geographical table invented at New Harmony whereby New York would be called Otke Notive).[21]

In the Education Society, the geologist Troost instructed in chemistry, the French artist LeSueur taught drawing, and Say organized natural history at the School for Adults for students aged twelve or older. The useful earth sciences were poorly taught as experimental farming by another Pestalozzian teacher Maclure had brought from Europe to New Harmony, William Phiquepal (also known as Phiquepal d'Arusmont). At the Infant School, the old soldier Neef used military exercises as part of the boys' instruction, for Maclure believed that gymnastics saved students from the "Diseases attached to sedentary habits." In his three-volume *Opinions on Various Subjects* printed at New Harmony, Maclure argued that useful subjects like natural history rather than "ornamental" learning would "promote the masses of producing classes," and he criticized Malthus for neglecting the economic efficacy of "educational reform."[22]

Maclure hoped to enroll eight hundred pupils in the schools of the Education Society, but he never got half that number. Most of the students were children of residents, and when Maclure began asking for tuition, enrollment dropped to about eighty. Meanwhile, Maclure and Owen argued. Maclure criticized Owen's management, and Owen questioned the abilities of Maclure's teachers.[23] Robert Dale Owen, caught in the middle, dropped out of the Education Society and returned to the vacant editorship of the *New Harmony Gazette*. His father dissolved his own share of the New Harmony communities on February 1, 1827, and left Indiana the following May. The New Harmony project had swallowed up four-fifths of Owen's fortune or $200,000, but Maclure, still determined to see through his educational aims, assumed the final property payments of $50,000 to the Rappites.

The duke of Saxe-Weimar Eisenach visited the community in April 1826. By his standards, "The living was frugal in the strictest sense, and in nowise pleased the elegant ladies with whom I dined."[24] The costumes worn by the people were odd looking. The women wore large pantaloons like those of little German girls, and the men wore wide pants buttoned over boys' collarless jackets.[25] The duke also visited the former house of George Rapp, a large, well-built residence, and noted: "The man of God, it appeared, took especial good care of himself; his house was by far the best in the place, surrounded by a garden with a flight of stone steps, and the only one furnished with lightning rods."[26] The duke remarked that in contrast, Owen lived modestly although he hoped to "remodel the world entirely," "to create similar views and similar wants, and in this way avoid all dissension and warfare."[27] Owen's ideas for land use within the community were certainly utopian. The duke noted that he would "only allow the public highways leading through the settlement to be enclosed," so that "the whole would bear a resemblance to a park." By the time a third royal visitor, Maximilian, Prince of Wied-Neuwied, spent the winter of 1832–33 at New Harmony, customs had changed. The inhabitants, now reduced to six hundred, dressed in "bad imitation of all the fashions of English towns."[28] The women sported large hats with loose veils and gaudy plaid mantles, which together "often have the most ludicrous effect in these remote forests." The men wore coats made of "common woollen horsecloths, white or green, with gay stripes on the collar, cuffs and pockets, nay, some are striped all over like zebras." The cost for such garments ranged from eight to ten dollars.

Owen's plans for a parklike development had not been realized. Whereas the Rappites were raising mulberry trees and fostering local silk production in Economy, Pennsylvania, the chief crop at New Harmony was Indian corn, and the principal diet was mush with milk. Living at New Harmony was too inexpensive to encourage good husbandry. Corn that fetched two dollars at the Canadian border sold for six and a half cents a bushel. Cattle and horses, given no hay in the winter because none was raised, ate the bark of the trees or starved to death. The community's pigs, fattened on the waste of a reopened whiskey distillery (built by Rapp but shut down by Owen) were allowed to die in the streets. Maximilian observed these carcasses "partly devoured by others." The carnivorous pigs of New Harmony had also shocked the duke, who noted that the once more numerous "rattlesnakes have a powerful enemy in the numerous hogs, belonging to the settlers, running about the woods, which are very well skilled in catching them by the neck and devouring them."[29]

After one year of testing at New Harmony, Maclure modified his educational ideas and shifted his full attention to orphans. On May 27, 1827, he announced the opening of the Orphans' Manual Training School. The founding of the School of Industry followed, and its "school sheet," the *New Harmony Disseminator,* was first printed on January 16, 1828.[30] Earlier Say had replied to a concerned parent: "Mr. Maclure's reasons for his change of plan, are the following: finding himself too old to contend any longer against the ignorance and deep rooted prejudice of parents he gives up all attempts at giving a rational education to those who can afford to pay for the education of their children, and returns to his original plan of giving knowledge cheap to the poor and oppressed."[31] Say concluded that "the inequality of knowledge is the source of all the evils that torment humanity." The Adult Society for Manual Instruction opened in 1828, but it soon closed. At the School of Industry, carpentry, woodturning, taxidermy, blacksmithing, cabinetmaking, hatmaking, shoemaking, dressmaking, and home economics were all taught. Apparently the instruction in drawing, printing, and engraving was so excellent that Benjamin Tappan wished to hire LeSueur as a private tutor for his children. Say responded: "As to Mr. Maclure's plan of published Books, he has always said, and yet says, that even should his other intentions be successfully counteracted, the business of Science shall go on."[32]

At New Harmony the "business of Science" was the production of books. By June 1830, Say was in charge of the *New Harmony Disseminator,* which he decided to use to publish his *American Conchology,* a work on mollusks, and during the winter of 1831–32, Say urged another loyal New Harmonite, Madame Marie Frétageot, to secure subscriptions while she was convalescing in Paris. Frétageot confessed to Maclure her doubts about the success of such a publication, for she did not feel that inexperienced students could produce a scientific work that met European publishing standards.[33]

Despite Say's failing health, the deterioration of the community around him, and the lack of such necessities as suitable paper and letterpress, Say pushed on with the *Conchology* and planned for the first part to appear in December 1830.[34] Say's wife, Lucy, drew the shells for sixty-six plates of the *Conchology* after her teacher LeSueur prepared drawings for two others.[35] Despite Frétageot's misgivings, Lucy Say colored 2,450 impressions with the help of two children, Henry Tiebout and his sister Caroline, who was paid two dollars a week for her work.[36] These children had come to New Harmony with Cornelius Tiebout, an outstanding engraver, who moved from Philadelphia to help Say execute the third and last volume of his *American Entomology*.[37] Maclure's idea of using the community facilities to produce a scientific work appealed to more than one New Harmony naturalist, and Say's colleague LeSueur published two parts of an awaited work on fish with colored plates and prepared a third for press. This work was never completed. Say declined to translate the rest of the text from the original French because for some unknown reason LeSueur refused to associate Say's name with the finished publication.[38]

If Maclure's exact political purposes are elusive, it is equally difficult to determine to what extent naturalists shared his social philosophy. In 1818, Nuttall honored Maclure with a newly established genus *Maclura,* which included the recently discovered western species, the Osage orange.[39] Later, he dedicated his supplement of Michaux's *Sylva* to his patron, but these were reasonable gestures of gratitude. Charles Willson Peale painted an affable portrait of Maclure for his museum in Philadelphia, and in 1824 his son Titian retraced parts of the earlier trek to Florida. A small picture of Owen is preserved in one group of Titian Peale's papers. Rafinesque visited the community at New Harmony and a utopian project at Valley Forge, Pennsylvania.[40] In 1833, he listed Maclure as one of the people who bought herbarium specimens from him.[41] These facts offer few clues to his true feelings, and none of these naturalists joined New Harmony despite their admiration for Maclure as a pillar of American science.

Whatever personal reservations he may have had, Say's correspondence indicates that he truly endorsed the goals of New Har-

Thomas Say (1787–1834), 1819, Academy of Natural Sciences. Shortly before the departure of the first Long Expedition, Charles Willson Peale painted Bartram's grand-nephew in uniform. In 1822, Say was appointed to the faculty of the Philadelphia Museum, where he worked with Titian Peale on his *American Entomology*. (Frick Art Reference Library)

mony's founders. Benjamin Coates claims in his carefully under-
stated obituary of Say that "in philosophy" Say was an advocate
for "that doctrine which attached exclusive importance to the evidence of the senses." "Fact alone was the object which he thought
worthy of his researches."[42] Say himself wrote to Harris that on
November 27, 1825, he left Philadelphia with a group of naturalists with the intention only of observing New Harmony during
the winter season, but, he continued, "Having arrived here, we
became interested in the singular spectacle of society which this
place presented, we were involved in the vortex of experiment to
realize the dreams of perfection in human association, which had
been so confidently & imposingly promulgated."[43]

Maclure and his urban followers were in many ways unprepared
for the squalid realities of frontier life. Troost, whom Maclure
described as well-meaning but too religious, quit New Harmony
for the new University of Tennessee, and even Maclure left the
community in 1828 for Mexico, where Say, who visited him in
1829, wrote that he hoped to achieve his goals untrammeled
among the Indians. Although Frétageot joined Maclure in Mexico,
where she died in 1833, Say stayed at New Harmony until his
death in 1834. The philosophy of Condillac and the teaching methods of Pestalozzi did not bring forth a "*second Adam*" or establishment of the natural system of classification at New Harmony, but
during the next decades, research in natural history which Say and
his colleagues developed through the Education Society prospered
in the frontier on a scale probably only Maclure imagined possible.

Maclure's Legacy

Maclure was childless, but the schools he financed produced a group of unusually effective educators, geologists, and political leaders. For example, from 1865 to 1873, LeSueur's former pupil Eli Todd Tappan served as president of Kenyon College, which has since maintained a distinctive approach to higher learning. Troost left New Harmony in the summer of 1827, but after he settled in Nashville he promoted geology at the new university. His student assistants included Abram Litton (1814–1901), who went on to study chemistry with Justus Liebig before joining Owen's son David Dale Owen (1807–60) on the federally supported survey of the Chippewa Land District, and Richard Owen Currey (1816–65), later the founding editor of three scientific journals. Lewis Troost, Gerard's son, was a civil engineer and wrote many newspaper accounts and popular leaflets to promote the railroads.[1] Maclure's emphasis on surveys, scientific journals, and inexpensive publications is apparent and, through Troost, reached other researchers as well. After visiting Troost for six weeks in 1832, Joseph Nicollet became interested in geological field work for the first time, and his results received the notable support of Louis Agassiz.[2] In the autumn of 1834, George William Featherstonhaugh spent two weeks with Troost en route to survey the Ozark Mountains for the federal government. Featherstonhaugh was a friend of Richard Harlan, now corresponding secretary of the academy, and Harlan may well have suggested this collaboration for, as Rafinesque bitterly complained, the well-known British-born editor had no qualifications whatsoever as a field researcher.[3]

The influence of the Education Society was not confined to geological surveys; it also extended to the founding of the greatest public resource for science in the United States, the Smithsonian Institution. In 1833, Owen's oldest son, Robert Dale Owen, returned to New Harmony and was elected to the House of Representatives from 1843 to 1847, the period when Congress began to take action on the $500,000 bequeathed to the United States by James Smithson. In the Senate, Benjamin Tappan of Ohio bolstered Owen's efforts in the House and voiced opposition to control of the bequest by the National Institution for the Promotion of Science, a private organization Joel Poinsett had founded in Washington to receive the collections sent back by the first U.S. Exploring Expedition under Charles Wilkes. The zealous abolitionist Tappan outlined instead a bill that emphasized the use of the bequest for practical scientific education. Tappan's bill was modified by Rufus Choate, senator from Massachusetts, who argued for a national library with an annual budget of $20,000 or two-thirds of the income from the bequest. The Tappan-Choate bill passed in the Senate but in the House Owen substituted a bill of his own, which cut spending for books to $12,000.[4] Owen's bill failed because it was opposed by John Quincy Adams, who as president had laid the crucial legislative foundation for the federally sponsored Wilkes Expedition. As George Ord had predicted from Philadelphia, the unprecedented bequest caused "an immense stir, a grand *speechification,* characterized by rant, fustian, and nonsense."[5]

Owen would not be deflected. Later in the summer of 1845, he drafted another bill that cut library spending for books to $5,000, and he proposed a building that reused the original and exceedingly ambitious architectural plans for New Harmony.[6] Arguing that "we must diffuse knowledge among men" and "not deal it out to scholars and students alone," Owen introduced another

Pestalozzian idea, a national normal school with graduate instruction for teachers.[7] Adams, of course, continued to oppose the idea of a federally supported school, and Owen's bill was soundly defeated. William Hough of New York, a friend of Owen's, redrafted a version, which passed in both the House and Senate. Hough's version, a compromise, included a board of regents, Tappan's idea, and also appeased Choate by increasing library spending by $5,000. Although his home state had in 1827 been the first to legislate the establishment of training institutes for teachers, Hough's draft did not include such a provision.[8] Led by George Perkins Marsh of Vermont, the library faction won in the House. Books, rather than museum collections or teachers, had come out on top at just the time when funds from Maclure's substantial bequest were being used to establish 160 workingman's libraries throughout the West.[9]

As a delegate to the state convention for the revision of the Indiana constitution in 1850, Owen fought for the extension of property rights for married women, the abolition of debtor's prison, and a state provision for public schools.[10] These reforms had all been intended as part of the way of life at New Harmony, but Owen's ideas did not prevail. Perhaps the voting public in Indiana associated women's rights with Owen's well-known agnosticism and advocacy of birth control. Owen's concern with the role of women in American society reflected his friendship with Frances Wright. In 1824, Wright and her sister Camille followed Lafayette's famous guest tour to America.[11] These young wards of Owen's mill partner Bentham were raised according "to his own peculiar crochets," and they soon demonstrated the unusual character of their upbringing.[12] Two years after their arrival in the United States, Frances founded the colony of Nashoba near Memphis to rehabilitate purchased slaves.[13] Soon afterward, at New Harmony, she met her future husband, Guillaume Sylvan Casimir Phiquepal d'Arusmont, and Robert Dale Owen, but, seeking an even greater audience, she left New Harmony for New York City in October of 1828. Within three months, Wright announced the publication of the *Free Enquirer,* a newspaper that opposed organized religion and advocated social reform. Owen and d'Arusmont joined her later that year and extended an invitation to Say to aid their enterprise as editor of the *Daily Sentinel*.[14] Even though his wife, Lucy, was from New York, Say declined their offer so that he could continue work on freshwater clams at New Harmony.

Maclure's goals met with greater success in the careers of Robert Dale Owen's brothers. Both the younger Owen sons became heads of large and important state surveys in the Midwest. After working on the Indiana survey in 1835, David Dale Owen was appointed state geologist in 1837. Two years later, as the new United States geologist for the Land Office, Owen moved the staff of mapmakers into buildings at New Harmony. Demonstrating his father's talents to mobilize people, Owen organized 139 assistants into twenty-four field parties that had by June of 1840 surveyed eleven thousand square miles of public lands in the old Northwest Territory (now Illinois, Iowa, and Wisconsin) for the Federal Land Commission and the Treasury Department. Because the Smithsonian bill had not established a national museum, the mineral specimens they collected were sent back to the National Institution in Washington, D.C.[15] In 1847, the indefatigable Owen surveyed the Chippewa Land District, and after 1852, he retired as state geologist of Kentucky and Arkansas. Owen's youngest son, Richard (1809–90), succeeded his brother as state geologist of Indiana, and as professor at the state university from 1864 to 1879 he was probably instrumental in diverting some of the valuable mineral collections housed at New Harmony to Bloomington.

By 1846, New Harmony had become a scientific stepping stone to the West in part because of the geological surveys and consequent real estate promotion and in part because of Maclure's bequest, which allowed his brother Alexander to live at New Harmony as a scientific middleman. Despite the loss of his house and library through fire, Alexander Maclure continued his deceased brother's efforts to attract scientific talent from the academy to New Harmony by offering parcels of land to persons such as Samuel Morton who had known the geologist.[16] Eminent foreign travelers such as Charles Lyell continued to visit the community, as well as notable regional naturalists such as Dr. George Engelmann of St. Louis and government professionals such as Leo Lesquereux, the remarkable paleobotanist who was totally deaf.[17] The classic memoir on extinct sloths which the academy's curator and most outstanding member, Joseph Leidy (1823–91), published in the *Smithsonian Contributions to Knowledge* relied heavily upon collections and notes David Dale Owen had made on the Kentucky survey. Furthermore, the paper that preceded Leidy's in the *Contributions* was an account of a tornado near New Harmony by Say's old friend, John Chappelsmith.[18] By midcentury, David Dale Owen's international reputation had established Say's goal, the community's scientific credibility at home and abroad.

Chappelsmith's skills as an engraver owed much to both Tiebout and LeSueur, and his interest in nature is evident in the large and beautiful plate of twisted trees that accompanied his *Contributions* article. On even a grander scale, the U.S. Geological Survey retained its mapmakers in the community's stone buildings until the completion of the Smithsonian Institution along lines which ironically reflected designs originally intended to replace the Rappite structures at New Harmony. Despite Gray's trenchant remarks about the community's old, worn-out letterpress, the New Harmony printer Josiah Warren is credited with the invention of the continuous roller press.[19] Warren had conducted the community's musicales and, after leaving New Harmony, established three utopian communities of his own. The most notable, now Brentwood, Long Island, was once called Modern Times, a name to be remembered fondly by fans of Charlie Chaplin's film classic of that title.

As a village apart from science, under Owen family management New Harmony became an extremely successful center for developing improved breeds of cattle, a reputation it enjoys to this day and an outcome of the utopian movement Jefferson, especially, would have applauded, although the private lives of its inhabitants were never fully homogenized into the mainstream of American custom.[20] Robert Owen and Robert Dale Owen remained confirmed spiritualists with an interest in phrenology. David Owen married Caroline Neef, the daughter of Maclure's Pestalozzian teacher, and shared their interests to the extent that he was willing to perform a postmortem dissection of his father-in-law's brain.[21] The youngest Owen brother, Richard, was found dead in 1890 after drinking embalming fluid.

Perhaps because of their peculiarities, one historian of New Harmony has doubted the value of scientists in politics.[22] The question is of particular interest not only for studies of New Harmony but also for the science of the period in general. Jefferson, of course, comes immediately to mind, and many other American-born naturalists had political ties of consequence. Jefferson's nephew Thomas Jefferson Randolph interned at Peale's museum in 1808 and 1809. Pickering's grandfather was an important statesman, and Say's father, Dr. Benjamin Say, was a state senator and twice congressman. Like Mitchill and Clinton, Torrey's father was active in New York politics and, after 1809, was appointed fiscal agent to the state prison, where Eaton was serving time for fraud.[23] Although the dramatic reversal of Eaton's career from felon to naturalist exemplified the moral reedification possible through the study of nature, Maclure, himself a public servant in 1803, did not specifically address Eaton's situation. His letters to Silliman repeatedly stated that science would fare better in the United States than anywhere else in the world because of its political character.

Maclure thought the progressive accumulation of scientific knowledge was the prerogative of a democratic society, and his leadership abilities, although impaired by illness and age, should not be underestimated. As he crisscrossed the frontier with hammer in hand, mapped and surveyed borders, and purchased obscure tracts of land, the geologist attracted an eclectic following for an intellectual revolution of his own agency. Was he influenced by another naturalist whom Jefferson had innocently encouraged, André Michaux, a traveling botanist who was also a French spy?

Maclure's program was intended to create citizenship in a new world bettered by social change. The "tabula rasa," Condillac's description of the human mind as a blank slate until marred by outside influences, the prejudices of the age, or other environmental factors, presumably explains Maclure's otherwise obscure insistence on achieving his revolution through the education of orphans.

He and Wright hoped that both racial and sexual differences would lose their import. Maclure seems to have shared Jefferson's view of the natural world as an awesome machine, the parts of which could and should be classified, but which did not change. The grand exception to his way of thinking was the parts of human society or the social classes, which were not to be confused with biological races or species. As recent events in the Americas, France, and Spain had clearly demonstrated to Maclure, the classes of society were

William Maclure (1765–1840). This oil portrait by Charles Willson Peale shows instruments the geologist used to complete the first geological survey of the United States in 1808. After 1817 Maclure served as the second president of the Academy of Natural Sciences in Philadelphia, and his interests shifted to educational reform. (Academy of Natural Sciences)

mutable and capable of revolution. Maclure believed that social classification of mankind had to change and evolve as men and women grouped and regrouped themselves according to the distribution of natural resources, wealth, and education.

Maclure envisioned many small societies, "separate yet federated," trafficking with each other in "the true spirit of equality" and "without injury to one another."[24] Is it coincidental that scientific authors using the presses he provided at the academy described true taxonomic groupings in the same way? Or that Rafinesque extended this thesis of mutability to all biological species? Taken together, their program of books, journals, large illustrated plates, and descriptions of thousands of species was made possible by Maclure, and like his social vision, their vision of nature appealed to readers seeking alternatives to foreign systems of thought.[25] Their work also shared another characteristic of Maclure's social projects—isolation. Say's manuscripts and collections were scattered between Philadelphia and New Harmony, and the ornithological materials Wilson, Bonaparte, and Nuttall described were dispersed after the public auction of Peale's museum. With the exception of Bonaparte, naturalists working in the circle of Maclure became increasingly detached from the mainstream of world science.[26] By 1842, the year that Gray became professor of natural history at Harvard College, Maclure and the other early academy naturalists were either dead or abroad, and Gray, taking a long view of American taxonomy, argued that their work must be ignored.[27] Thaddeus Harris queried, "But what can be done in regard to Le Sueur? who seems forever lost to his friends & to the world, after a debut the most brilliant. Will he too pass away without leaving behind any memorials of his eventful career, or any one to record the history of his life and labors?"[28]

Maximilian recalled pleasant hours spent leafing through LeSueur's portfolio, "a large collection of drawings and descriptions," and he noted LeSueur's "specimens, for the most part stuffed," which the French naturalist sent back to the national museum at Paris rather than to the academy in Philadelphia.[29] The prince clearly hoped that LeSueur would return to France, and after several extended field trips (one with Troost in 1831), LeSueur did leave the New Harmony community permanently in 1836.[30] Two years later, George Ord visited him in Paris, where he was teaching painting again, and by the time of his death in 1846, he was working as the curator of a natural history museum in Le Havre. LeSueur did not realize his full potential, and he never produced what Maximilian, an accomplished taxonomist himself, recognized as "an important supplement to the work of Cuvier and Valenciennes."[31]

Other than Say's *Conchology,* little is known about the scientific projects carried out at New Harmony. Owen had originally planned to partition off the ends of a cross-shaped building (the design for which Rapp had thrice received in a dream) for use as a library, natural history cabinet, and schoolrooms, and Owen's museum idea was developed to some extent. The granary fortress built by Rapp was used to house Maclure's extensive mineral collection cataloged by David Dale Owen. The quality of the drawings in a young person's botanical notebook found among the Owen family papers indicates a high level of instruction, probably by LeSueur, but as he told Say, he was not irrevocably committed to American science. Say, in contrast, appeared content to remain at New Harmony. He apparently desired few material comforts and was described as "quite comical in the costume of the society, with his hands covered with hard lumps and blisters, occasioned by the unusual labor he was obliged to undertake in the garden."[32] When the wage system was reinstituted in 1830, Frétageot informed Maclure that Say had declined a salary. Say did demand a first-rate scientific library, which Maclure provided, and even after Say's death, Maclure directed 2,259 scientific volumes to the community.

The academy reclaimed many of these books and benefited from some other New Harmony projects as well.[33] Maclure had purchased the engraved copperplates for Michaux's *Histoire des arbres forestiers de l'Amerique septentrionale* in Paris, and after his death, the executor of his will, his brother Alexander, sent the plates to Morton from the academy, where they were used along with lithographic plates to illustrate editions of the *North American Sylva* supplemented by Nuttall.[34] In France, Maclure also purchased the valuable Audebert-Vieillot plates of ornithology, which

Maximilian saw exhibited in the library at New Harmony.[35] This ambitious international ornithology was never produced at New Harmony. Although the French work would have been an important step away from the strictly American focus characteristic of academy studies, from a practical point of view there was no need for this large bird book. Wilson's *Ornithology* was being republished in several inexpensive British editions, and Audubon's volumes were in press.

For Maclure, the pursuit of scientific knowledge was a moral exercise in which natural history had a special place.[36] He had viewed the mill schools of New Lanark as large, carefully executed experiments with human subjects. He saw cheap publications as the means of advancing the working classes, but did he also expect these outlines and epitomes of some of the same expensive books he had financed at the academy to advance science? Say, his wife, and their New Harmony helpers prepared the *Conchology* in a format Say had developed on the pages of the academy's *Journal* published before he became involved in the New Harmony movement. Thus Say's participation in the community did not alter the way he organized or presented data. It did give his terse style a social justification that was lost on his distant and increasingly critical colleagues at the academy.[37] Say did not have to publish the *American Conchology* as a serial in the community's newspaper. He deliberately chose this format so that the shells "may be readily recognized even by those who have not extensive cabinets for comparison."[38] He knew his primary audience was not the experts who had access to the collections he had initiated in Philadelphia.

As the announcement in the *New Harmony Disseminator* promised, the purpose of Say's work was to serve the frontier as a substitute for a reference collection. Although he was eager for European subscriptions to the *Conchology,* Say did not ignore the needs of his frontier readers. His glossary to the *Conchology* published in 1832 explained terms used and gave the "General Rules Relative to Diminutive Compounds" based upon the standard authorities as well as "Rules for the Pronunciation of the Linnaean Names."[39]

Despite Say's youthful declaration that American conchology

required a natural method of classification, the method he used for the sixty-seven genera given in the *New Harmony Disseminator* was based upon the shell, not the animal inside.[40] This artificial method annoyed at least one contemporary with technical interests, C. A. Poulson, who argued that Say should study "the animals more, and the shells less."[41] In fact, Say did not overlook the animal's physiology. He discussed such subjects as the habits of *Petricola,* which, as its name implied, bored into calcarious rock, and the locomotion of the snail *Glandia.* Poulson was the translator of Rafinesque's *Monograph of the Bivalve Shells of the River Ohio,* a work originally published in Brussels in French, but his complaint that Say had ignored his client's descriptions of seventy-five species, forty varieties, and twenty-eight figures sounds suspiciously like Rafinesque himself.[42] Before he began publishing the *Conchology,* Say had written Rafinesque's onetime colleague at Transylvania University, Charles Wilkins Short (1794–1863), requesting "all such works by Mr. Rafinesque as include description of new N. American animals, and particularly his treatise on Unio." He continued, "The latter work," the *Monograph,* is especially desirable "for quotation," and true to his word, the text to plate 53, *Unio quadrulus,* mentions Rafinesque's nomenclature.[43]

Had Poulson perhaps forgotten that he (exhibiting Rafinesque's characteristic vanity) had presented Say with an autograph copy of the *Monograph* in 1820? This copy, now owned by the American Museum of Natural History in New York City, is replete with Say's marginalia, which give corrections and synonyms for Rafinesque's nomenclatural designations. The letter to Short provides some insight into Say's working methods and conditions at New Harmony: "I have not seen that work Rafinesque's 1820 'Monographie' since my departure from Philada; by the copy of it belonging to the Academy of Nat. Sc. I labled [*sic*] many of the specimens which were then in my collection, but in removing them to this place many of the lables were misplaced, & I cannot therefore rely implicitly on any of them." Only six issues of the *Conchology* appeared before Say's death on October 10, 1834. The seventh issue was edited by T. A. Conrad, who copied the descriptions for plate 65 and most likely for plate 64 from William Turton's *Bivalves of the British Islands*. Conrad obviously did not believe that the American species were unique, and he did not share Say's opinion of Turton so that the last issue lamely relied upon a British author Say did not esteem.

Several years after her husband's death, Lucy Say left New Harmony to live with her sister in New York, and on Harris's request the academy forwarded Say's insect collections to his care in Boston.[44] In a letter to Harris, Lucy Say expressed her desire to learn engraving to continue the *Conchology,* and within her limited means (Maclure left her an annuity of $3,000), Lucy Say intended to carry on her husband's work and to reintroduce his New Harmony writings to the eastern scientific establishment.[45] In 1858, her efforts came to fruition when W. G. Binney edited Say's complete writings on the conchology of the United States, and the next year, John L. LeConte published Say's entomological work.[46]

Unfortunately, Harris died before LeConte's edition was completed. His hopes for a permanent position in natural history at Harvard having been frustrated by Gray's appointment, Harris continued as librarian of the college and was unable to reorganize Say's materials, which arrived in great disarray. More than a decade elapsed, but Louis Agassiz, who had acquired a few of Say's specimens with the Melsheimer collection, encouraged LeConte to procure new paratypes for species Say had collected on his trip to Mexico with Maclure.[47] In an effort the old surveyor Maclure would have applauded, LeConte's study of the distribution of California beetles in 1851 produced the first map of faunal regions of the western United States, and LeConte has been called the greatest American student of Coleoptera. His father, Torrey's childhood friend John E. LeConte, had prevented the publication of Say's papers in the lyceum's journal in 1825, but on the eve of Darwin's *Origin of Species,* LeConte's 1859 edition was reprinted in 1876, 1885, and 1891.[48] After 1844, Agassiz, as is well known, tried to centralize American zoology at Cambridge, Massachusetts, by acquiring collections from all parts of the country. At this time, Gray did not appreciate the "remarkable proportion of endemism [that] characterizes every portion of western North America," and when Nuttall, for example, designated small concrete genera in his *Sylva,*

Gray tended to retain larger comprehensive ones.[49] In contrast, Agassiz canvassed among both amateur and professional naturalists and reevaluated the taxonomic contributions of earlier (often self-taught) explorers including Say, LeSueur, and Rafinesque. His preference for splitting systematic groups in part reflects their work and is a neglected aspect of his rivalry with Gray, which as their biographers note preceded their famous debates about evolution.[50]

PART FOUR

Publications

The Rise of Peer Review   10

Following the War of 1812, the most active naturalists in this country were descendants of the eighteenth-century Enlightenment. Despite the undeniably religious impulse behind the establishment of eighteenth-century botanic gardens and other horticultural pursuits in this country, zoologists rooted in Jeffersonian thought, if not openly critical of religion, remained at least silent on theological matters. In 1819, Titian Peale noted that on the banks of the Ohio River the Long Expedition party was visited "by a commissioner of the Bible Society who left us two bibles to remind us in the wilderness that we carry the prejudice of____."[1] Titian's colleagues preferred to view the wilderness without the "prejudice" of organized religion. After 1825 chemist Thomas Cooper echoed Maclure when he called for "full and unlimited freedom of examination in every department of knowledge," and freethinker Frances Wright continued to demand unfettered inquiry "into the nature of things" before crowds in every major city across the country.[2] Their words challenged the American Protestant intelligentsia, who, fearing an inherent social and spiritual chaos in their message, wished not to abolish scientific inquiry but to merge it with Christian natural philosophy and a sound framework of religious belief.

The prominent role clergymen played in British natural history had no counterpart in the United States until well after 1842, when

125

a generation of American-born Protestants led by Torrey, Gray, and James Dwight Dana achieved scientific leadership.[3] Even with the mounting conservatism that accompanied Gray's and Dana's professional ascent, it was their work and not the social consciousness Owen or Maclure hoped for that gave American natural history a firm methodological foundation. Furthermore, their assertion of religious questions in science led to fruitful intellectual debate about the nature of the species—their origins and numbers—that climaxed in the Darwinian Revolution. Ord characterized the early Academy of Natural Sciences as a "club of humorists," and it is true that by 1842 Wilson's ideal of natural history was no longer practical.[4] During his early years, Say is said to have slept under the mounted skeleton of a horse at the academy, but as time went by, the poverty of field naturalists who had been the most active became increasingly apparent and embarrassing to resident members. Nuttall was criticized for renting a room from a Negro oysterman, and in 1838, another colleague, disappointed that Nuttall sold a collection "for the sum of ten dollars," unfairly commented, "Avarice, however, has always proved superior to friendship."[5] Science at the academy had outgrown clubhouse rules.

Linnaeus had originally intended his system of classification to provide a vehicle of communication, a universal language for natural history that would overcome national boundaries, but at the academy classification proved to be an unpleasantly disruptive element, dividing the scientific community its patriotic character was intended to unite. By 1842, the major controversies involving academy members arose from the use of new nomenclature.[6] The vicious tone of these nomenclatural debates is startling. Harlan and Godman, for example, divided their Philadelphia colleagues over the dentition of a new rodent and the classification of the extinct *Mastodon giganteum*.[7] In violation of the basic tenets of the academy's publication committee, Harlan vilified Godman in an unsigned piece charging that "like the envenomed asp that has missed its aim," Godman would hide his own "head in its hateful coil."[8] In cryptic letters to Isaac Hays, Godman referred to Harlan as "Dr.

Jellybag," or "Zoologicus," who "I believe will P———ss on the Academy at length."[9]

Despite the vulgar language and personal attacks, advocates of new nomenclature were far less distressed about their mistakes than were their critics. They realized all too well that current conditions of field work made their achievements highly susceptible to error. This is not to imply that they were careless. Their concern with minute anatomical distinctions suggests the opposite. As Say wrote to Thaddeus Harris in 1825, "We are all liable to commit errors, but very glaring ones have a tendency to cheapen the character of American science."[10] In the field, Say, Nuttall, or Peale had to accept limitations of accuracy as long as there was no easy way to transport reference libraries or collections to the western outposts where they worked and no secure means of sending back large numbers of specimens or adequate samples of the same species. Even east of the Mississippi, Say had to ship collections from Indiana to Boston via a middleman in New Orleans.

The American scientific community recognized the handicaps field naturalists encountered and responded by discouraging their publications. Critics also recognized that the taxonomy used by field authors for insects, mollusks, mammals, and plants continued to exemplify Wilson's approach to classification some thirty years after the celebrated ornithologist's death. Established groups were subdivided, and variant forms were classified as new "allied" species. Of course, such "refinement" not only generated novel material for publication but also led to error and synonymity, which many naturalists wished to end. Early academy members had welcomed Maclure's generous gifts of books, equipment, and money for travel and building without which the academy's expanding publications would not have been possible. Maclure also wanted the academy collections opened to the public on a regular basis, but by the mid-1830s the resident membership would not accept this demand from their long-absent president. For twenty-three years Maclure made his ideas known to the academy primarily through his correspondence with Morton, the organization's longtime secretary and historian, but living in Mexico after

1827, Maclure was unknown to many younger members.[11] Of those academy members who had directly benefited from his generosity, only Titian Peale remained in Philadelphia after 1826, and the interests of city members drifted away from Maclure's expressed goals.

Organizations like the academy or the New-York Lyceum of Natural History discouraged technical publications by field naturalists in various ways. In the initial stages of projects, professional courtesies were often denied them, and access to reference collections, in some cases their own specimens, was restricted. Periodical editors refused to continue publishing their works as the status of the exploring naturalists on western expeditions was redefined. Most effectively, the review process of works that were published excluded field authors.

Although American and European members of the scientific press welcomed the second volume of Say's illustrated *American Entomology,* Say soon encountered trouble publishing unillustrated papers in the *Annals of the New-York Lyceum.* DeKay wrote Bonaparte that the head of the editorial committee had "repeatedly stated" that he would "defy anyone to understand Mr. Say's descriptions" without plates. DeKay, the lyceum's president, said that he found Say's work "minute" enough, but he was loath to overrule the committee. Say, now at New Harmony, had left manuscripts with both the lyceum and the academy for publication, but having received no direct communication, he responded, "What can all these societies be about? They seem to be slumbering."[12] Say was able to prepare the third volume of the *American Entomology* for press only because a friend "had the goodness to offer to attend to the printing" in Philadelphia.[13] By 1830, the silence of Say's scientific peers seemed complete. He had neither seen the third volume after it was published nor received "any reviews or heard any remarks whatsoever respecting it."[14]

More seriously, field naturalists complained that their use of reference collections they had helped to establish was being restricted. Members were understandably cautious about the security of valuable scientific materials forwarded to New Harmony, but Say concluded that his publication was being effectively obstructed by the lack of common professional courtesy. "I am possibly to find in the same *safe keeping* several other boxes and books, that have been sent to me to Philadelphia," he wrote, "which I cannot as yet get account of."[15] Nuttall became similarly indignant that he should be denied access to Torrey's herbarium even though Torrey had complete freedom to study Nuttall's numerous specimens in the academy's collection. "I was, I had thought," he wrote to Asa Gray, "entitled to expect, the same privilege of consulting Dr. Torrey's heb.^m. that you have of consulting the herb. of the Academy, tho it is now determined, I find, that I shall be obliged to work in the *dark*."[16] Furthermore, in Townsend's absence, the academy gave Audubon unprecedented access to the undescribed bird skins Townsend had sent back from Oregon. An academy committee prepared a paper on twelve of the new species to ensure Townsend's priority, but this was an ineffectual measure after the sale of ninety-three bird skins, which furnished Audubon's *Birds of America* with seventy figures.[17]

When Townsend traveled to Oregon with Nathaniel Wyeth and his party of "Bostons," he was, like Nuttall, in the joint employ of the American Philosophical Society and the Academy of Natural Sciences. From his journal, it is fair to conclude that he saw his role on the Wyeth trip as comparable to Titian Peale's assistantship to Say on the first Long Expedition, and in addition to birds, the young ornithologist collected mammals, insects, and Indian skulls.[18] During the fifteen years since the Long Expedition, however, the professional standing of field naturalists had changed. Gray increasingly considered them to be hired collectors, not technical authors of their own discoveries, and even Silliman ceased to recommend their works to inquiring readers.[19] The upshot of this trend was that field naturalists, like Townsend, faced slim prospects when they attempted to publish their finds, and coming after Audubon's sumptuous fourth volume, Townsend's *Ornithology of the United States* was, despite the scope indicated by its title, reduced to twelve pages and one plate.[20] The status of academy collections was so undefined that Harris, fearing Townsend might reclaim his Oregon collections of insects, urged Morton to use his authority so that he, Harris, could publish them.[21]

Nuttall responded to this shabby treatment of his western traveling companion by allowing the junior naturalist to publish the popular account of their travels or tradebook under his name.[22] Meanwhile, Nuttall, an established botanical authority, returned east to discover that his own western findings sent back to Torrey and Gray for publication in their proposed *Flora* had been ignored or reorganized beyond his recognition. Nuttall's only recourse was an appeal "in consideration of what I had done, not in the closet but in the field."[23] Townsend had hoped to follow a pattern of supplementation established in American ornithology by Wilson, Ord, and Bonaparte. Nuttall had also worked with Wilson's collections at the Peale museum, and in the 1830s, he, too, attempted to carry this precedent into botany with a supplement to Michaux's *Sylva*. Even though this work proposed to illustrate the western trees Nuttall had recorded for the academy on the Wyeth trek, the *Sylva* would not have materialized in print without the provisions of Maclure's will. Nuttall saw the writing on the wall and having received a modest inheritance from his family, sailed for England after Christmas of 1841.

Philadelphia colleagues also redefined their role with respect to federally funded expeditions, a major employer of field naturalists and an important outlet for their taxonomic efforts. Before Titian Peale went west with the Long Expedition in 1819, he received instructions from both the government and his father to collect and review representative western fauna first observed by Lewis and Clark. The 122 drawings he made in the field were copied by others but never published, and although DeKay and Clinton agitated for better knowledge of the mammals of North America, there is no evidence that they encouraged Peale to publish.[24] When Peale sailed home with the United States Exploring Expedition in 1842, he was only forty-two years of age, but the creative period of his career as a naturalist was over. Having been in and out of the country for the better part of a decade, Peale was not part of the scientific milieu to which he returned after four years at sea, and, as is well documented, his crowning achievement, the official report on the birds and mammals of the expedition, was suppressed.[25] Unlike the other members of the scientific corps, Peale

Titian Ramsay Peale (1799–1885). Peale painted this self-portrait with help from his older brother Rembrandt, an accomplished history painter who encouraged Titian's field studies of western Indians although his interests lay with butterflies and birds. (American Museum of Natural History)

was unable to surmount the considerable delays of government bureaucracy. The expedition's commander, Charles Wilkes, was a young, ambitious lieutenant with limited experience at sea, and friction developed between naval officers and the "Scientific Corps of Civilians," particularly Joseph Pitty Couthouy (1808–64), an amateur conchologist from Boston, who became disgruntled with curtailed freedoms and limited opportunities to go ashore to collect.[26] Wilkes required that the scientific journals be surrendered to him periodically for examination.[27] This somewhat jealous demand discouraged liberal expression on the part of the naturalists but enabled Wilkes to follow the direction and progress of the civilian scientific work. Had Wilkes used his authority wisely, this interim inspection could have circumvented the prolonged disorganization that characterized the publication of the scientific reports for the next thirty years.[28] Echoing the sentiments of the academy's first publishing committee, the Library Committee of Congress decided to publish only what was new and authorized an edition of one hundred copies for university and library purchase (although at one time there was talk of giving every senator a copy). By 1849, the estimated publication costs were $89,370.

In contrast, Congress allotted only $5,000 for the care of the collections. Peale complained of vandalism, and in 1842 Congress appropriated the more realistic sum of $20,000. The natural history specimens were moved to the upper room of the Patent Office "under the care of such persons as appointed by the Joint Committee on the Library," and although Peale applied for the appointment, Pickering was made curator.[29] Members of the scientific corps, officially confined to Washington until the completion of their reports, still found themselves separated from reference collections. For Peale, the consequences of this isolation were disastrous as he spent six frustrating years working on the expedition's collection of 762 specimens of 38 mammals and 265 birds. Peale's *Mammalogy and Ornithology,* for which he received $6,840, was issued in 1848, but ten years later, the government sanctioned a second report on the birds and mammals by John Cassin, who had not accompanied the expedition and did not consult Peale. This second report was accompanied by a large, illustrated atlas which included plates based upon Peale's elaborate drawings. Originally,

Peale was asked for plates of 15 mammals and 69 birds, but this complete atlas was never published. By the time he was dismissed from the scientific corps in 1848, Peale had completed only 46 of the 84 plates.[30]

American ornithology, once the acme of the American field tradition, was no longer the bastion of the exploring naturalist. Despite Peale's personal role in the development of the academy's superior bird collections, members of the academy gave Cassin full institutional support. Echoing Nuttall, Peale's defense of his "rights" carried little scientific weight: "Cassin's report and mine are in existence . . . the world may judge between the rights of original observation and closet philosophy—what an observer says, and what others think he ought to say."[31] Peale pursued an undistinguished career in the U.S. Patent Office. He explained to Harris, who many years earlier had also lost out to Pickering for a job, "A series of misfortunes have conspired to prevent a professional connection with the pursuit of Natural history, but never have I at any time lost my early love for its allurements in the field." Peale was unable to reconcile the changed status of the field naturalist within organized American natural history. He wrote to Harris, "The great disappointment has been in finding so few persons, who like you have a kindred feeling."[32] Peale's situation with regard to the Wilkes reports was not unique. Couthouy's contributions were also described by an outside expert, Augustus A. Gould, who had not accompanied the expedition. Couthouy sailed with the *Vincennes,* Wilkes's ship, and Peale, assigned to the *Peacock,* was apparently unaware of the full extent of the younger man's tribulations. Like Couthouy, the more experienced Peale was self-taught, and Couthouy's eventual dismissal included all the elements of his own situation. At best, Wilkes viewed Couthouy (who was himself a merchant ship captain) as a troublesome amateur.

Cassin and Peale were working within the same framework of pre-Darwinian science, and since profound changes in the American study of ornithology did not occur until after 1908, many of the differences their peers perceived in their work have become obscured. Peale's success was jeopardized at the start by the dual nature of his assignment—to study both birds and mammals. Ornithology was so well established a science in America that by the 1840s, the academy housed world-class collections of bird skins. In contrast, the study of North American mammals had not made parallel progress for two reasons. Peale's early plates of western mammals remained unpublished, and mammalogy, far more than ornithology, relied upon skeletal evidence and internal anatomy rather than skins. The long Pacific voyage of the U.S. Exploring Expedition squadron precluded optimal preservation of complete specimens, and many specimens were confusingly mislabeled on arrival or damaged in storage.

Sanctioned by the Library Committee of Congress, the circumstances surrounding Peale's suppression were far more dramatic than other field naturalists' situations, but the consequences for the American careers of all were the same. Like Peale, Townsend went west as an expedition assistant in 1834, but upon his return, he discovered that field experience alone could no longer assure his qualifications as a taxonomist or his right to publish his own discoveries. Nuttall's botanical career was curtailed by many of the same difficulties, and Say's association with New Harmony alienated him from the circles of science and learning his work required. All these naturalists received what Peale aptly termed "a cold shoulder," and increased government control of expedition collections served to widen the distance between exploring naturalists and the institutions they represented. Although "the conduct of certain men in office in Washington with regard to our friend Peale" came to the attention of the American Philosophical Society in 1842, the "outrage," it seemed, was "unknown" to the members.[33] Four years later, Ord concluded that he did not know what the "fate" of "Poor Titian" would be.[34] By the mid-1830s, however, Ord and his colleagues were committed to narrowing, not broadening, the base of scientific participation.[35] Publications were key, but just because Titian Peale or George Ord initially sought the same means to achieve their goals does not imply that their goals continued to be similar or even compatible. When Bonaparte or Gray began to travel abroad, they strove to raise the standards of American science, not by rivaling European works but by associating their aims with European achievements.

The Business of Science

In his *Opinions on Various Subjects,* Maclure had pronounced that the "best, most useful and cheapest pastime is the natural sciences," for "they banish envy and promote contentment."[1] These were well-meaning words coming from an affluent businessman, but the excessive speculation that followed the War of 1812 led to the crisis of 1819, a year that found every young naturalist associated with the Academy of Natural Sciences seeking employment on western expeditions.[2] Stephen Harriman Long (1784–1864) was an early contributor to the academy's collections, and from 1816 to 1823 he led a series of expeditions into the American interior which were the only significant overland army explorations between the War of 1812 and the 1840s.[3] After 1818, Torrey, Nuttall, Say, Titian Peale, Rafinesque, Keating, and possibly LeSueur all sought for a place on Long's topographical expeditions.

Between 1809 and 1829, the national per capita income declined by 20 percent, and times were hard for young men who wished to pursue natural history as a full-time profession. It is not surprising then that when Robert Owen visited the academy with ideas for a western community that offered scientific livelihood, Say, LeSueur, Chappelsmith, Speakman, Troost, and later Tiebout left Philadelphia to take up residence at New Harmony.[4] New Harmony provided the possibility of employment with a modicum of the comforts absent from western travel, although as Rafinesque noted, the appeal of Maclure's "views and fine Colleges" was not of long duration.[5]

Life on the frontier was neither content nor cheap, especially if game was scarce. Titian Peale noted that in Missouri the hungry Long Expedition party paid a Shawnee Indian $1.50 for a deer, a third more than white hunters charged. At St. Charles, Missouri, he wrote, an inhospitable settler "would neither sell or give us anything, and granted us even water with bad grace." Supper that night consisted of one of Peale's specimens, a hawk, and Say became ill from "long fasting." At Kennedy's fort, they were able to purchase a ham for ten cents a pound and, for another twenty-five cents, a loaf of cornbread, milk, and grain for the horse. A week later the *Western Engineer* anchored at Cote des sans Dessein to clean the boiler. The French-speaking town subsisted by hunting, and young Peale replaced his army-issue fatigue coat with "a leather hunting shirt for 4 dollars and a pair of leggings for $2.50."[6] Soon afterward, Peale developed "excruciating" pains in the soles of his feet, a serious problem because the steamboat proved impractical and the naturalists had to walk most of the way upriver.[7] At Franklin, Missouri, William Baldwin, the expedition's botanist, died, and during the following spring, scurvy and exposure claimed 160 men of the army contingent that reconnoitered with Long's party near Council Bluffs.[8]

Western travel could indeed be grim, but the exact role of frontier experiences upon subsequent relationships with the resident scientific community is difficult to determine. After they returned from their travels, explorers were unable to pursue successful scientific careers in this country, and the New Harmony group did not rejoin the society of the eastern seaboard in which they had developed their interest in natural history. Their descriptions of landscapes, Indians, or even new species failed to substantiate conventional ideals about nature, and what scientific insights they gained on the frontier remained outside the mainstream of natural history. Ten years after Maclure led the "boat-load of knowledge" to New Harmony, Rafinesque's views on the "Pleasures and Duties of Wealth" made no inroads for him in the scientific establishment.[9] After Wilkes terminated Peale's appointment with the scientific corps of the U.S. Exploring Expedition in 1847, the academy withheld support until 1872.[10] The few biographical details recorded for Townsend similarly suggest unfullfilled promise. His family, early advocates of prison reform, sent young Townsend to Quaker schools, but although he studied dentistry (a field that had also intrigued Peale's aging father) he never practiced. After the Wyeth trek and illness in South America, Townsend returned to Philadelphia in 1838 to serve intermittently as the academy's curator in 1839–40 and again in 1845–46.[11] He spent part of the interim working in Washington, D.C., on the bird collections of the Wilkes Expedition, and he may have collaborated with Peale on specimens obtained in Chile. Townsend died in 1851 before a scheduled voyage around the the Cape of Good Hope.

The botanist Edwin James fared no better. Soon after the publication of his account of the first Long Expedition, James was appointed surgeon of the United States Army and accepted a station at Fort Crawford (Prairie du Chien). Thereafter, he became increasingly interested in philology and, in 1833, translated the New Testament from Hebrew Chaldaic into Chippewa. Three years later, James moved his family farther west, where, in the words of his biographer, "his view inclined to ultraism."[12] James and his son completed a survey of six townships in Illinois and Missouri in 1844, but, sharing a common fate with his former cohort, Titian Ramsay Peale, James was not paid what he had been promised, and his contributions were suppressed. He passed the remainder of his life outside the confines of organized science and became fanatically committed to the cause of Indian temperance. The most disinterested reader can only wonder to what extent contact with the American frontier shaped the intellectual horizons of naturalists working with western collections. By 1847, even Bonaparte, much to his American friends' dismay, had become leader of a radical republican party in Rome.[13]

In 1838, the Corps of Topographical Engineers was created by an act of Congress. This event, following the panic of 1837, was disastrous for naturalists of the West, who could not compete with federal surveys. At best, their subsistence as collectors with specimens for sale had always been precarious, and unlike medical botanists, self-taught zoologists, no longer young, could not qualify for appointments as army surgeons. Nuttall and LeSueur, finding no occupation open to them, left the United States permanently. With the legal establishment of a defined group of experts responsible for systematic surveys and land classification, the role of the naturalist on western geographical expeditions became, if not obsolete, at least supernumerary. After 1838, eastern colleagues could avail themselves of technical geographical and meteorological records maintained by the Corps of Topographical Engineers. They did not need to read or rely upon someone else's field notes, and Gray and Torrey found their biological specimens could be gathered less expensively by less accomplished collectors than Nuttall. With rapidly escalating costs of publication, Rafinesque's advertisement of a century of unique specimens for $50 suggests that he found it cheaper to furnish the actual specimens for sale than to draw and prepare plates for publication.[14] Furthermore, as the roster of scientific personnel chosen for the Wilkes Expedition shows, field experience was no longer necessary for appointment to a military expedition.[15] In 1838, Peale and the other naturalists who accompanied the Wilkes Expedition received $2,500 a year plus provisions for what proved to be a four-year commitment. Peale was later paid $6,840 for finishing his expedition volume and Cassin was paid $2,999.93 for its replacement, but Gray and Agassiz, who like Cassin had not accompanied the expedition, were paid a combined total of $11,316.66 to produce four volumes, only one of which materialized.[16]

By 1838, the year the expedition set sail, the actions of Congress had effectively ended any economic incentive for field naturalists. In addition, although the state surveys of the 1830s focused upon natural history and geology, naturalists who had modeled their careers on Wilson's could not, like Harris, settle for a regional perspective. Finally, despite Jefferson's earlier encouragement, the zoological branches of natural history were the last scientific divisions to find acceptance in the college curriculum, and academic positions were scarce.[17] Unfortunately, full-time jobs often included administrative duties that were not compatible with field work. Charles Orr expressed a shared concern that Say's scientific abilities had been restricted by his increasingly heavy administrative load at New Harmony, for within a month of his arrival, the entomologist had been elected superintendent of literature, science, and education, a position he filled until his death.[18] Critics who did not travel could not have any real idea of the conditions that compromised western investigations. In 1838, Rafinesque was horrified to learn that Torrey had unthinkingly burned grasses collected by John Bradbury from the territory of the Mandan Indians.[19] In Oregon, Townsend's entire collection of snakes and lizards was destroyed by a drunken fort tailor after he guzzled the whiskey from the two-gallon carboy in which they were stored.[20] One hungry soldier on the Long Expedition complained of having to use water contaminated with buffalo dung for soup, and another compared the highly flavored, if dubiously preserved, rations to the qualities of a cigar.[21]

Say had described Maclure's view of books as the "business of Science," and one of the most important problems faced by all of the early nineteenth-century field naturalists was that of financing their work. As Nuttall, whose original desire to come to the United States stemmed in part from reading F. A. Michaux's book of travels, attempted to explain to Gray, field naturalists could not be expected to support their own research. He described his domestic situation with candor: "I am obliged to use the utmost economy to live."[22] Commenting further on the relationship of collector to collections, he added that "one sett" of dried plants to the academy "is as much as I owe a country that never patronised or assisted me in anything and to explore wh. I have sacrificed much property."[23] Because of his British citizenship, Nuttall was excluded from any formal participation in military expeditions, and having been rejected from the Long Expedition in 1819, he was forced to fend for himself that year in the dangerous Arkansas Territory.

The expense of publications was central to Nuttall's problems, and the correspondence of Rafinesque, who bravely pursued an American career despite all the impediments posed by the scientific community, is full of complaints about publications.[24] He even described for Jefferson "a new kind of Western Literary Inquisition" in the small university town of Lexington, Kentucky.[25] Although Ord seemed surprised at the apparent "plot" afoot to "defraud" participants of the Wilkes Expedition of "the produce of their labors," Peale, for one, was already familiar with the tribulations of illustrated publications. In 1833, his proposed work on the Lepidoptera of North America had not proved feasible in large part because his method of illustration, copperplate engraving, was too expensive.[26] Peale later referred to this project as a "labor of a life," and his inability to publish his artistic studies, which incorporated the work of his deceased brother, remained a lifelong frustration.[27]

Costs of illustrated publications were rising on both sides of the Atlantic. Like their fellow Scotsman, Maclure, the Chamber Brothers of Edinburgh devoted their book business to the democratization of knowledge by printing inexpensive tracts. Although fortunes were to be made from cheap literature, in 1826 many publishing houses failed in Great Britain.[28] Subsequently, with the notable exception of Audubon's elephant folios, the British market favored modestly priced serials and pocketbooks. In 1827, the publisher Archibald Constable issued the first volume of the *Miscellany,* which still bears his name, although he died that same year. In 1831, Constable's *Miscellany* issued an illustrated, five-volume edition of Wilson's *American Ornithology,* which was followed in 1832 by Sir William Jardine's pocketbook edition of the same work.[29] Dr. Dionysius Lardner's *Cabinet Cyclopedia,* with essays by William Swainson, and William Home Lizar's *Naturalist's Library* are other examples. These works of natural history, which sold from five to six thousand copies, were financially viable because the illustrations were more likely than not pirated; to avoid artists' fees, plates were traced from other works and sketches were stockpiled for future use.[30]

Samuel Augustus Mitchell had published volumes written by Say and Bonaparte and illustrated by Titian Peale in 1825, but apparently cognizant of the British situation by 1827, he refused to continue Say's *Entomology.* He reluctantly agreed to publish a final third volume if the edition was reduced to fifty impressions. Thereafter, Say seriously entertained the idea of a New Harmony School Press edition similar to Constable's *Miscellany,* which had already incorporated the volume on birds he had helped Bonaparte write.[31] Mitchell also discontinued Bonaparte's *Ornithology.* For the second volume, the young prince turned to Carey, Lea, and Carey after looking unsuccessfully for a publisher in New York City.[32] Henry Carey felt "very indifferent about publishing the remainder of the *Ornithology*" because profits from such elaborate volumes were "very small," and sympathetic naturalists regretted that a "bookseller" who held that much "power" did not use it more "liberally."[33] The engraver retained for the second volume was Wilson's old friend Lawson, but the artist was not Titian Peale, who had collected most of the specimens and illustrated the first volume, but Rider, one of Wilson's colorists. The reason for this change is not known, but William Cooper, who supervised the volume, remained on good terms with Peale and consulted with him from time to time.[34] Progress on the plates for the fourth and last volume remained at a standstill until Lawson completed illustrations his daughters colored.

Because American authors often used different systems of classification, illustrations were essential to the success of any work. In addition, illustrations functioned like type specimens if preservation of type materials was not possible.[35] The commitment to illustrations, despite their cost, also established the author's credibility within the scientific community. For example, in 1825 Harlan's *Fauna Americana* scooped Godman's *American Natural History,* but because Harlan's book was not illustrated, it did not receive the critical attention it deserved. Not to be outdone, Godman's publishers, Carey and Lea, with "a most entire contempt" and "full of despise" for Harlan, mustered the resources of Peale's museum to illustrate Godman's three volumes published the next year.[36]

Harlan and Godman were professional rivals, and their vicious debates extended from the practice and teaching of medicine and natural history to the doors of Peale's museum.[37] The uncanny similarity of their texts reveals both their similar scientific goals and the limited resources available on North American mammals. Godman was (or pretended to be) less interested in new species and was able to salvage his manuscript with illustrations of well-known animals, most of which were exhibited at Peale's museum.[38] The plates that made his book more popular than Harlan's were engraved from earlier drawings made by Rider, LeSueur, and Charles Willson Peale, but although Godman, like Harlan, relied heavily on Long Expedition materials, he used none of Titian Peale's western drawings.[39] This omission is particularly surprising because Godman had married the daughter of Titian's half-brother, and Godman and Titian Peale had collaborated in the field.[40] The finished plates based on Rider's drawings, however, resemble Peale's studies and may be tracings. Since LeSueur was in New Harmony, the plates that bear his name presumably were those executed for a work Ord initiated after Barton's death in 1815, but did not complete, and which were then retained by Lawson for later use.[41] Similarly, Charles Willson Peale's plate of the fossil bison that appeared in Godman's third volume of 1828 must have been drawn for another purpose long before the octogenarian's eyesight began to fail.[42] That the illustrations for Godman's volumes were not colored represented a substantial savings for the publishers, but after his author's fee was drastically reduced, Godman concluded that he had been obliged "to sell myself" to a bookseller. Despite Godman's cryptic comment that Titian Peale had some "observations" to share on the matter, Peale's reaction to Godman's exploitation of his family connections is not known.[43]

Godman, a skilled translator, aided the French-speaking Bonaparte in the extensive corrections of Wilson's taxonomy which were published separately from the illustrated supplements to the *American Ornithology*. He did so in secret for he recognized the jealous protection of the ornithological press by Ord, who he wrote only half-jokingly "I suppose is terrified."[44] His fears were not ill-founded. Ord had revised a facsimile edition of Wilson's *American Ornithology* at great personal expense. He hired Helen and Malvina Lawson to color the plates, using Peale's mounted birds as guides, and over a period of three years he paid them

$5,000, "at that time an almost unheard sum for two young girls to earn."[45] Having made this investment, he could not welcome activities that made his edition of Wilson's work immediately obsolete, and he was understandably reluctant to share the market with either Bonaparte's supplement priced at $180 or the corrected addenda sold at the New-York Lyceum of Natural History for $3.50. As the academy's vice-president from 1816 to 1834, Ord used his position to force Audubon to publish abroad.[46] Given Ord's power and the limited press capable of high-quality publications in the United States, we can now understand, if not condone, Godman's secrecy and his undetected (and unnecessary) efforts to rig the reviews of his popular *American Natural History* by writing the copy that appeared under others' names.[47]

Having undergone a recent religious conversion from French materialism, Godman is the likely translator of Duke Bernhardt's *Travels*. In the American edition and translation a footnote was added to the original text, which condemned New Harmony life as "the promiscuous intercourse of sexes and colours, the downfall of religion, and the removal of all restraints imposed by virtue and morality."[48] Ord shared this image of New Harmony, and the obituary of Say which he read before the academy in 1834 was considered so critical of Say's and Maclure's involvement that he was asked to resign his vice-presidency.[49] A kinder biography was substituted in the press, but many members endorsed Ord's objections to New Harmony if not his acerbic public expression of them.[50] A few months before Say's death, Philadelphia had witnessed the first of a series of riots in which white mobs attacked black neighborhoods.[51] Although some naturalists at the academy, particularly the liberal Quakers, may have sympathized with blacks' aspirations for upward mobility, the footnote to Bernhardt's *Travels* shows their estimate of the appropriate limits was conservative.

Ord's letters continued to discuss social problems with correspondents, and despite the academy's reprimand, Ord maintained the esteem of his colleagues. In December 1851, he was elected president of the academy and held the office through 1858. A man of learning and ability, Ord had in his youth compiled data for the first edition of Webster's dictionary which he felt were used without proper credit.[52] Acknowledgment became Ord's sore point, and throughout his career he championed the cause of Wilson in an effort to discredit Audubon.[53] He was able to demonstrate that Audubon, in the not uncommon trade book practice, had traced illustrations from Wilson's volumes which, ironically, Ord's edition had made more widely available.[54] In one of his least choleric descriptions, he dismissed the *Birds of America* as "les mille et une nuits de l'histoire naturelle."[55] Although he was touchy about proper acknowledgment, Ord did not see the same issue in Say's situation at New Harmony, and after Say discontinued translation of LeSueur's manuscripts, Ord took up the task, which "remained in embryo."[56]

Just before he died, Say remarked that "the delay of my Philada friends, of several years, has been of serious inconvenience to me," and the deterioration of Say's relationship with the academy, specifically with Ord and Pickering, was indicated by the institution's unwillingness to publish his work in the journal whose standards he had established in 1817. Having been turned down by both the New-York Lyceum and the academy, Say turned to the *Contributions to the Maclurean Lyceum,* the organ of a satellite group within the academy membership loyal to Maclure after 1826. Say was their president in absentia, and Bonaparte was the vice-president, but publication in the *Contributions* was not assured. In 1833, Say complained to Morton that his descriptions of Hymenoptera which had been printed in the third number had ended abruptly with the phrase "to be continued." The paper was not continued, and Say wanted an explanation.[57] Say also questioned the editing of his manuscripts by Pickering, the academy's new curator and librarian.[58] Say had first exchanged specimens with Pickering in 1825, when the latter was a young man residing in Salem, Massachusetts. After Pickering was hired as the academy's librarian, he served as member of the consulting committee for the Wilkes Expedition. Pickering also controlled access to collections Nuttall and Townsend sent back from Oregon, and it was Pickering who allowed Audubon to examine the bird skins for purchase without Townsend's knowledge.

As Maclure feared, the academy was implementing measures in his absence which effectively separated the Philadelphia community from the New Harmony "reformation." When Maclure offered a substantial gift of money to the academy on the condition that the annual dues be lowered from ten dollars, his patronage was refused for the first time. After Say's death, Pickering went to New Harmony to retrieve the valuable library for the academy, but the scientific publications Maclure made possible at the community were denigrated. Reviewing the edition of Michaux's *Sylva,* Gray sarcastically queried: "Mr. Maclure's munificent intentions are—fulfilled, shall we say? by the edition now before us, printed at New Harmony, Indiana, upon wretched, flimsy, white-brown paper, the wornout type, which seems to have done long service in the columns of a country newspaper."[59] In truth, the *Sylva* was no less attractive than Gray's *Flora* published with Torrey. Gray was simply reiterating Ord's characterization of Say's *Conchology* as "a disgrace," a book possessing "repulsive character."[60] Indeed, after Ord's suppressed obituary of Say, Lucy Say was forced to offer for remainder half of the entire edition of the last volume of the *American Entomology* for three dollars a copy.[61]

Ord's views have dominated later evaluations of New Harmony in large part because his obituary was given a second life when LeConte used it as an introduction to the second volume of Say's collected writings. In 1890, E. A. Schwarz, a beetle specialist for the United States Department of Agriculture, wrote in a frequently cited passage, that "to read between the lines," Say had no other place to publish than "the most obscure village paper that could be found in this country."[62] Ord's exclusion of Audubon from the American scene was less successful, even though only eight members of the academy subscribed to *Birds of America* as compared to twenty-six from the American Philosophical Society. As is well known, Audubon came to Philadelphia in 1824 to look for a publisher and a patron. He visited the Academy of Natural Sciences, and he also visited the studio where Titian Peale was drawing birds for Bonaparte's supplement to Wilson's volumes. The young prince immediately acknowledged his talent, and Audubon's first published plate illustrating the "great crow blackbird" or the purple grackle appeared next to Peale's work in the 1825 volume. Although Bonaparte honored his commitment to Peale for the rest of the first volume at least, in Europe, he lent his uncle's famous name to the so-called "Bonaparte" subscription list which Audubon used to sell his books abroad.[63]

Ord's actions may have contributed to Wilson's lasting popularity, but they also had some negative repercussions for the very institution he wished to preserve. After touring North America with the Swiss artist Karl Bodmer, Maximilian did not offer his important collections and notes to the academy. He took the plants and land animals back to Europe for description by Nees von Essenbeck at Breslau and Achille Valenciennes, Cuvier's co-worker and director of the zoological museum in Paris.[64] He left the birds at the New-York Lyceum, which had welcomed Audubon and published the final corrections of Wilson.[65] Maximilian's decision suggests that he, at least, was impressed with Audubon's abilities and considered no European more competent to describe American birds.

Ord had argued for standards of quality, and during the period that followed the War of 1812 those standards were as important for publishing as for research. The two went hand in hand, and as the number of publications increased, they replaced the correspondence that had directed scientific activities early in the century. At the same time, the status of full-time researchers such as the naturalists who had joined Owen's utopian community was changing. Despite his admiration for Wilson, Ord bitterly criticized Say for his total commitment to natural history. Ord was a prosperous merchant, but his wealth did not lead to personal happiness. His wife was committed to an insane asylum, and his son, perhaps in a pale imitation of Wilson, pursued an undistinguished career as a painter. Such pressing concerns at a time when phrenologists instructed the public that improved habits could alter social station no doubt caused Ord to be skeptical of Owen's psychological interests.[66] Ord concluded that "it seems by no means advisable" to recommend natural history to the early attention of youth, "lest what was intended merely for pastime become an occupation."[67] Maclure, by contrast, advocated natural history, especially geology, as crucial for youth and the future of a republican democracy. Both men, despite their different social philosophies, devoted their time and much of their financial resources to the same "business of Science," books, and in opposite ways they sought to control American natural history.

Anyone who doubts the early strength of organized American science must witness the power of its censure. In 1817, Rafinesque was praised before a natural history class at Williams College as a naturalist second only to Torrey, and his writing credentials were so esteemed that he was made a member of the lyceum's publications committee.[68] After 1819, Rafinesque was forced to publish privately or in obscure magazines. He remained unemployed for the better part of his life, and after his death, his collections were heedlessly destroyed as "trash."[69] Gray spoke for a school of American botanists when he described Rafinesque's nomenclature as "fictitious, and unworthy of the slightest notice."[70] Rafinesque's extreme ideas about the constitution of species had nothing to do with Gray's severe evaluation of his work. At a time when Rafinesque was still regarded as a naturalist in good standing, he had entertained ideas about permutation which were, by his own admission, far more radical than anything he later wrote.[71] Rafinesque's contemporaries remembered him instead for his taxonomic blunders such as the nine species of fish he naively described on the basis of sketches "drawn from nature" by a then unknown river merchant named John James Audubon. Had Rafinesque's English been better, he might have guessed that Audubon's new species of "Devil-Jack Diamond fish" might just as well have been "its own Genus" of buttermilk flapjack.

After 1817, Rafinesque was not blameless for the adverse reception of his nomenclature in the United States. His sharply worded criticisms of "aristarchs and moles" provoked rapid response, and his surpassing talent for vituperation did little to make amends.[72] In 1819, the *American Journal of Science* refused to accept his manuscripts, and the academy declared his work "unworthy of publication," "the wild effusions of a literary madman."[73] Despite earlier encouragement from Mitchill, Rafinesque soon discovered that he could not interest scientific publishers in his work.[74] Initially impressed by Rafinesque's knowledge, DeKay came to dread his "neologomania" and "miserable hawking after new species."[75] Eaton wrote that the women of upstate New York adorn their "witticisms" with the name of Rafinesque, and of course Audubon's use of Rafinesque as the model for his hilarious "eccentric naturalist" in the popular *Ornithological Biography* did not enhance Rafinesque's reputation. Even Say hoped that "for the honor" of natural history, "all naturalists" would join in "the determination to pay no attention" to Rafinesque.[76] All the same, Rafinesque would not be silenced or ignored, and he continued to deluge his friends and foes alike with missives and printed pamphlets. In 1832 alone, a year of relative affluence for him because of patent medicine sales, his postage bill amounted to $23.45, more than one-third of his annual rent.[77]

Just before he moved to New Harmony, Say described what he thought was the greatest scientific obstacle of his time, the "scarcity of good Zoological plates in this country in consequence of the high price of the books which contain them."[78] Having identified the problems of cost and scarcity, he continued, "More therefore seems to depend upon the acumen of the student." Perhaps with Maclure's plan in mind, he added, "Placed at the commencement of an era in which, judging from our present prospects, natural history will receive a portion of attention corresponding with its importance, it will be vain to suppose that our observations will always be free from error." But error was not the real issue. The rise of peer review, reaction to Owen's "eccentric experiment," and controlled access to reference materials and the press were all factors working against field naturalists during the 1830s. As a group, their reputations, nomenclature, and personal collections (often discarded, neglected, or dispersed by auction) did not survive the decade. In 1842, Gray simply regarded Rafinesque's ideas as the last, most extreme expression of a taxonomic style which American systematists had already rejected in the works of Nuttall, Eaton, and Say.

On hearing of Rafinesque's death in 1840, Gray wrote to Silliman that he was trying to persuade Torrey to prepare a "notice," but Torrey declined, having endured almost a quarter of a century of Rafinesque's unsolicited correspondence.[79] Gray decided to undertake the chore himself, but it proved more onerous than he anticipated, and within five weeks the notice had grown to "an article of several pages." Gray even traveled to Philadelphia to track down obscure publications.[80] By New Year's Day, the review had reached "15 to 18 pages," and Gray felt it was important enough "to be the first article in the issue" of the *American Journal of Science*.[81] An equally thorough examination of Rafinesque's zoological works was prepared by the Philadelphia naturalist S. S. Haldeman. Both reviews were lengthy and harsh. Neither reviewer gave much attention to Rafinesque's theory of permutation so his unique approach to American taxonomy was lost as his systematics were lumped together with the works of Nuttall and Eaton, authors Rafinesque had been the first to criticize in 1817.

In 1833, Rafinesque had complained that his travels alone cost him between $8,000 and $10,000, but of the sixty-seven hundred binomials he proposed for his finds only thirty generic names are recognized today.[82] As a great ichthyologist of the late nineteenth century noted, Rafinesque was uncontrolled "by the influence of other writers," or that "incredulous conservatism," which furnishes "salutary balance to enthusiastic workers."[83] Clearly, the recommendation of his *Flora Telluriana,* "Imitate my zeal, and be happy in the lovely study of flowers," was not upon his own example good advice for "enthusiastic" naturalists. In hindsight, a better motto might have been, "tread softly," the common name for Rafinesque's spurge nettle, *Bivonea stimulosa* (Michaux) Raf. Say had hoped that the "fault will not be ours if we should be deceived by bad descriptions or by the exhibition of unessential characters" in the works of European authors. His former colleagues, however, were neither so kind nor so confident in their position "at the commencement" of what Maclure called the "science of society," and they would not tolerate alternatives to the normal route for scientific publication, peer review, and publication committees.

Science by Management 12

In 1828, Richard Harlan lamented that "men of natural science are scarce here."[1] This observation was an acknowledgment that the New Harmony movement had taken its toll upon academy membership. The leaders were no longer in Philadelphia, and the city's learned societies were "too accustomed to enlarge their libraries by presents and begging"—from Maclure, he might have added. Harlan also commented that his colleagues, in marked contrast to his correspondent Audubon, were doing "very little" in the way of "scientific publication at present." The reason he gave was the expense of the requisite books. After 1817, Maclure had supported libraries and a wide spectrum of activities ranging from organized inquiry at the academy and the Geological Society of America to his rural school programs in Delaware County, Spain, and Indiana. Although he objected to the elitism of the academy, what Rafinesque called science "by management," he did not discontinue his considerable support of the institution, and he had the great wisdom to insist that library collections be housed in a fireproof structure.[2] Similarly, he did not abandon New Harmony after his disenchantment with Owen's motives or after his best teachers either died or quit. Indeed, Maclure's patronage became such an integral part of American natural history that it created its own system of checks and balances with the academy on one side and New Harmony on the other.

Progress often requires intellectual friction, and nowhere is the efficacy of this principle more evident than in the history of biology in the United States. By 1842, Gray had no opposition within the natural history community, but at this time even he recognized that no great ideas, no *"second Adam,"* were forthcoming. The arrival of Louis Agassiz in the United States changed this unproductive peace, and the establishment of zoology and natural history in the college curriculum during the 1840s owes as much to Agassiz's charismatic public lectures as to any indigenous development within American education. The brilliant Swiss naturalist immigrated to the United States on C. L. Bonaparte's encouragement.[3] Once established at Harvard, he boldly sought wide support for his Museum of Comparative Zoology at just the time Peale's museum was closed. Agassiz also revived a tradition of American natural history which stemmed from Peale and Jefferson to Wilson and his followers at the academy in Philadelphia. He acquired what was left of their collections, and his opposition to Gray often represented a more learned expression of their ideas about the uniqueness of species, including human groups, in the New World.

The study of race with its reliance upon anatomy, geography, languages, history, and religion had by midcentury become the most complex branch of zoology, and the science of ethnology had particular relevance to a country where by anyone's count there were at least three different human types. By 1842, the year Albert Gallatin founded the American Ethnological Society, the spectrum of feelings included three positions in addition to Bachman's proslavery religious stand.[4] The Jefferson-Morton position later championed by Agassiz emphasized physical parameters and present geographical location as the basis for classification. This approach lent itself to the idea of separate regional creations. The Barton-Mitchill position, which relied on Buffon's land bridge and linguistical studies, maintained that the Indian cultures of the New World were not the result of slow racial progress but were the remnants of a civilization diffused from Asia. In the extreme, this view regarded present-day hunter-gatherers not as the noble savages Jefferson described but as degenerates. A third position was more optimistic because it was not comparative. Like Maclure or Rafinesque, the advocates of this approach, best termed developmental, did not recognize the white European as the ideal type or epitome of creation: "Whence we ought to love each other whatever our shape, bulk and hue, as brothers of a single great family." Rafinesque noted that "man is a variable being, like every other, and subject to the EXTERNAL DIVINE LAW OF PERPETUAL CHANGE AND MUTATION; in form, size and complexion."[5] In 1835 he was scheduled to deliver a series of lectures on the subject of man at the Rensselaer Institute. Eaton described the event for a local newspaper: "'*Shades of variety*' as *evidence of specific differences* comprise all his supposed heresies."[6] The benign tone of Eaton's notice and the place of its publication suggest that there were Americans eager to hear a discussion of developmental theories even if they pertained to man.

All of the positions on human speciation assumed much more geological time than the biblical account of creation allowed, and each contained a valuable kernel of scientific truth. Separate creations, the pluralism of the physical approach, required an implicit acceptance of the fact that individual species can arise independently if they are geographically isolated. Old World origins—migration from Asia—emphasized populations, not individual parameters, and necessitated an understanding of the exploitation of new habitats. Developmentalism attempted to remove the connotations of degeneracy from descent and to regard adaptation in a way that did not confuse cultural values with biological needs. The slug was not inferior to the snail because it lacked a shell nor were the "esquimaux" lower than the Europeans because they lacked the wheel. Yet none of these arguments any more than Bachman's could explain why the orangutan possessed a voice box but could not speak. Although Pickering, unlike the followers of Condillac, minimized the adaptation of language as merely a "matter of convenience," other naturalists required a theory of speciation that explained the physiological basis for speech.

Gray praised Pickering's writings in print, but as a systematic botanist, he did not share Pickering's compulsion to discuss human biology.[7] Time had made an important difference. In 1800, political and scientific rhetoric was permeated with optimism for the human race, and people believed in racial improvement made possible by a democratic form of government. By the 1840s, failure to assimilate the Indian tribes and the continued enslavement of blacks led to pessimism in many quarters and a belief in their racial weakness.[8] Until about 1830, most American thinkers viewed the Indians as potentially equal with whites, and their savage state was not perceived as permanent. The justification of Maclure's Indian projects was intervention and the modification of surroundings so that Indian orphans could be reinstated at the level of white American society. A decade later, Morton's *Crania Americana* seemed to provide hard data for those who rejected Maclure's environmental approach as too Lamarckian and who were attracted by the idea of separate creations.

Darwin's *Origin of Species* would emphasize the role of specific competition, rather than environmental factors, in the determination of a given species' territory. Nuttall had arrived at this conclusion by 1832 in his discussion of the bald eagle, but ironically he was unable to give it general application for animals or his specialty, plants.[9] Rafinesque had considered the idea of limited competition in plants when he observed that introduced plants "appear to invade the fields and drive out the native plants in some instances," but despite the obvious parallel to settlers and Indians, he did not suggest competition among animals as a factor in speciation.[10] Pickering, in contrast, fully acknowledged competition but considered it to be an artifact of human migration and introduced species. Therefore, he believed that competitive "invaders" deemed to be native species were in fact naturalized weeds.

As noted earlier, the published sources for Rafinesque's ideas—Adanson, Necker, Lamarck, and DeCandolle—were all available to other academy readers, and evidently other academy members were willing to consider some of the same problems in a less provocative way. In 1832, Nuttall modified Lamarck's discussion of the sloth to include an American bird, the avocet, again in reaction to Buffon's theory of degeneration, and in 1836, Harlan weighed the comparative merits of Lamarck's progressive chain of being.[11] In his response to Buffon's classification of birds, Bonaparte had by 1827 become familiar with the theories of Vigors and Swanson, which Rafinesque claimed influenced him before

1815.[12] Although the quinary system did not necessarily imply evolution or even permutation, these ideas were compatible, as Robert Chambers demonstrated in 1844 in his controversial *Vestiges of the Natural History of Creation,* an anonymously published work which contained an illustration by LeSueur.[13] Any method of classification that favored the designation of varieties as new species was only a short step away from viewing permutation as the means of speciation, and having just completed a lengthy overview of Rafinesque's botanical writings, Gray recognized the theological ramifications of Chambers's *Vestiges.* He remarked that "we must receive" scientific truth cautiously "if proven and build up our religious belief by its side as well as we may."[14] For Gray, then, the permutation of species became a question of proof. Although Gray applied new critical standards to the work of his American contemporaries, developments in British natural history did not render earlier American efforts immediately obsolete. In his notice of Nuttall's supplement to Michaux's *Sylva,* Gray praised Nuttall's "quickness of eye" and "tenacity of memory."[15] He added: "In some of these respects, perhaps, he may have been equalled by Rafinesque,—and there are obvious points of resemblance between the later writing of the two, which might tempt us to continue the parallel." Since Gray had recently published a severely critical review of Rafinesque, Nuttall, by this time in England, had little to gain by this comparison. He also had little to lose. In his supplement to Michaux's *Sylva,* which was not printed until some time after his departure, Nuttall had described a member of the natural order *Torreya* as a "stately evergreen" discovered by H. B. Croom and "frequently called stinking cedar."[16] The history of the published names proposed by other field naturalists as well bears out Agassiz's later observation that "there is a world of meaning hidden under our zoological and botanical nomenclature, known only to those who are intimately acquainted with the annals of scientific life in its social as well as its professional aspect."[17]

Certainly the social and professional aspects of American scientific life in the early nineteenth century deserve closer attention, particularly since academy explorers brought together elements of frontier learning and its institutional counterpart. Arthur Bestor has noted the American tendency during this period to enact social reforms by founding communitarian societies rather than through legislation. On the scientific level, the exodus of academy naturalists to Owen's project at New Harmony is an example. Once at New Harmony, these naturalists found themselves in disagreement with certain of Owen's ideas, and they reacted by forming a group of their own, the Education Society. In Philadelphia, academy naturalists who were sympathetic to their views founded the Maclurean Lyceum of Natural History in 1826. The formation of the academy itself in 1812 with its strongly worded manifestos declaring intellectual "independence" was an early example of this same tendency, reflected by 1817 in the editorial policy of their journal. Only what was believed to be new was published, to the exclusion of revised or reworked materials and reviews. Ironically, this dictum meant that although the academy's journal provided a record of new species and discoveries, the much needed revisions of old groups could not appear upon its pages, and revisers like Bonaparte were forced to publish elsewhere.

Wilsonian-styled zoologists hoped to implement a rapid conversion of their science from the artificial to a natural system of classification. In contrast, botanists Torrey and Gray envisioned a more stately process of reorganization without any promise of novelty, the primary motivation of all exploring naturalists be they zoologists like Say or botanists like Nuttall. With the notable exceptions of Rafinesque, Nuttall, and eventually Eaton, early nineteenth-century botanists retained middle-class ties to the medical profession and horticulture, to the values of social protocol and real estate. Even William Bartram's "flowery sayings," like his father's earlier account of Georgia and Florida, were deliberately used to promote white settlement in the Southeast.[18] After William retired to his gardens in 1777, there were no roaming types like Alexander Wilson among the accepted ranks of botanists in Philadelphia or elsewhere, and no botanist left the comfortable environs of Philadelphia to join Owen's social movement in Indiana. The field explorers anticipated "revolution" in their beautiful publications, but the early sensitivity of Torrey to the international aspects of systematic biology made botany rather than zoology the real seat of taxonomic reform, and during the next decade botany emerged in the United States as a stable but singularly unoriginal science.

By 1836, the year Rafinesque outlined his natural system in his privately printed *Flora Telluriana,* most American naturalists shared the same goal—a natural method for the classification of species. Field naturalists continued to be interested in published discoveries, whereas Torrey, Gray, and their followers, most of whom had not traveled farther west than Buffalo, New York, waited for a revolution to happen within European natural philosophy. Their expectations were high. These Protestant botanists hoped that the true natural method would reunite natural history with the biblical account of creation and wrest systematics away from both the agnostic approach taken by early members of the Academy of Natural Sciences and the materialism of its president, William Maclure. Until that time, they resigned themselves to the limitations of the Baconian method of natural history: "All that we can do is to throw our pebbles upon the heap, which shall hereafter, when they become sufficiently accumulated, become the landmark of Systematic Botany."[19]

Torrey began the onerous task of cataloging specimens sent back by government expeditions ranging from John C. Frémont's exploits in the Great Basin to the gargantuan railroad surveys of the midcentury. Unlike any of the books inspired by Alexander Wilson's work, these massive reports were difficult to read. Hooker rightly complained of them to Gray: "Who on earth is to keep in their heads or quote such a medley of books double-paged, double-titled & half finished as your Govt. vomits periodically into the great ocean of Scientific bibliography."[20] Gray was well aware that American natural history was floundering like a rudderless ship in a vast sea, but without a comprehensive theory of speciation, the descriptions of large collections could not be handled with the ease and elegance which Wilson had envisioned possible five years after the Louisiana Purchase. Gray identified his role in the United States as a revisionist to reunite natural history and organized religion, and after 1859, natural history did function as an "appendix to the bible" and not even the best one as both intellectual biblical exegesis and religious fundamentalism gained increasingly strong holds on American thought.[21]

When Gray traveled west for the first time in 1877, he wore a suit and collected plants from a railroad car which carried his wife, a cook, a picnic table with a white tablecloth, and other accoutrements of civilized life.[22] He traveled in comfort and in the company of one of the world's greatest authorities on plants, Darwin's close friend Joseph Dalton Hooker. Standing in the midst of the Rockies, Gray shared nothing with field naturalists such as Nuttall, who had preceded him on foot some sixty-five years earlier. By the same token, their only living representative, Titian Peale, now an old man residing in Philadelphia, had absolutely nothing to say about Darwinism and still hoped for the day when the federal government would purchase his entomological collections of a lifetime.[23] The "revolution" Nuttall had predicted in 1818 occurred in the same year as his death, in 1859, and in his native land, Great Britain, and once again American thinkers attempted to find social meaning for biological principles in an intellectual effort aptly termed social Darwinism.[24] As Swainson remarked, "The revolutions of science are almost as frequent, and often more extraordinary, than those of political institutions."[25] Maclure would have agreed with Swainson that "it is, therefore, the part of the natural not less than of the political historian, to trace the causes of such revolutions." Once again, western expansion, the plight of the Indians, and Negro education became issues of focus, and after the rediscovery of Gregor Mendel's work in 1900, the perfection of man reemerged as an achievable possibility under the banner of the eugenics movement. At the academy, researchers attempted to incorporate the agrarian vision of Jefferson with the educational reforms of Maclure in a way that would make real the role for stable collections first perceived by Charles Willson Peale. Their failure should not be surprising, for as Rafinesque observed, employment and the opportunities to establish a reputation were created by management—the federal government, state surveys, and scientific societies—and not by ideals.

In 1824, Maclure wrote to Silliman that American intellectual progress was impeded by what he called "colonial bias" or imitation of the English. Actually, early nineteenth-century American natural history did not lack distinct goals or a "pure science ideal."[26] Although the British Linnaean Society had been founded in 1788, unlike the Academy of Natural Sciences, it did not publish a journal until 1856. The Zoological Society of London was established in 1826, fourteen years after the Philadelphia academy, and its proceedings date from 1830.[27] Like Ord and Maclure at the academy, most of the members of the Zoological Society were professional or businessmen in practice or in retirement in the city. The growing number of "country" naturalists, however, was soon reflected in the need for a committee of science and correspondence.[28] Despite its unusually early commitment to a journal, the Philadelphia academy was less successful in recruitment, and during the 1830s, the number of American naturalists never justified the creation of societies centered around special disciplines like the Royal Entomological Society founded in 1833 or the British Botanical Society established in 1836. All the same, Maclure's schools and the New Harmony project passed early academy goals to a younger generation, and natural history captured a larger public through the pictures painted by the Peale family and through illustrated publications based upon materials at the Philadelphia Museum.

The Peaceable Kingdom   13

Maclure had associated the development of government with climate. In his view, the warmth of the tropics had given rise to tyranny, whereas cool air was conducive to freedom. Even "the priests," he concluded, "the precursors of all tyranny," could not exist in Lapland. Maclure argued that Europe offered the most favorable climate for "the perfection of the human species"; the United States was still in an experimental stage. Climate also played a role in his plan of education which stressed the importance of botany. "Plants," Maclure claimed, "as they are freer from bad habits than animals, may be easier accustomed to any climate."[1] Properly acclimated plants also avoided the expense of hothouses, an important point, for Maclure also realized that all school programs for "Industrious Producers" could not have the benefit of his patronage. In this regard, Maclure's thought was not exceptional. From the inception of Penn's "holy experiment" in Pennsylvania, botany and horticulture had enjoyed the benefits of Quaker prosperity. William Penn was a determined and adventuresome gardener, and during the eighteenth century, an outstanding circle of Quakers established the illustrated study of plants in the British colonies. This group originally included James Logan (Penn's secretary), the British merchant Peter Collinson, John Bartram, Joseph Brientnall, and later, Humphrey Marshall, John Fothergill, and of course, William Bartram, who, in 1803, illustrated *Elements of Botany,* the first textbook of its kind to be published in this country and a work that went through six editions by 1836.[2]

A highly successful farmer, John Bartram is thought to have organized the first American water-use cooperative, and in the early nineteenth century, the Society of Friends continued to emphasize the efficacy of conserving resources, proper land management, and hard work.[3] Such Quaker optimism offered a timely antidote to the discouragement Buffon's theory of degeneration put before New World hopefuls immigrating to the United States after the French Revolution. Indeed, the happy association of Quaker values with American plants persists to this day on the package of the breakfast cereal Quaker Oats. Conscious promotion was not needed to assure European attention to the botanical riches of the New World. The organization of botanic gardens incorporated the Linnaean system, but these gardens were also ornamental, and according to the taste of the times, they presented visitors with a pleasing medley of plants. William Bartram's descriptions of the Florida wilderness are perfused with this same curious bifurcation. The Linnaean method offers a beauty of order; the plants allow indulgence of sense: "What a beautiful display of vegetation is here before me! Seemingly unlimited in extent and variety; how the dew-drops twinkle and play upon the sight, trembling on the tips of the lucid, green savanna, sparkling as the gem that flames on the turban of the Eastern prince; see the pearly tears rolling off the buds of expanding Granadilla; behold the azure fields of cerulean Ixea!"[4]

Although Bartram's recurrent comparison of East Florida's "natural" gardens with "paradise" or the restoration of Eden is extreme, the religious impulse motivating the creation of gardens in Great Britain or among the colonial Quakers should not be ignored. On each side of the Atlantic, however, there were important differences. As Oliver Goldsmith noted, the construction of large parks typical of eighteenth-century English efforts necessitated social engineering. Entire villages were removed and roads diverted to create them.[5] In sparsely settled America, these social costs could be avoided. Indeed, the biological wealth of the new nation thoroughly Americanized the Edenic vision of the paintings of Edward Hicks, a devoted Quaker who resided in Bucks County, Pennsylvania.

Although little is known about his methods of working, Hicks's compositions of the *Peaceable Kingdom* are probably the most famous examples of nineteenth-century folk art. Despite their undeniably religious purpose, they provide a unique pictorial record of the popularization of American natural history over a period of thirty years. By 1824, when Owen first recruited at the Academy of Natural Sciences, the seeds of utopianism which found visual expression in Hicks's earliest paintings were already present in the Philadelphia scientific community, and Maclure was already known to the art world for his patronage of John Vanderlyn in Italy.[6] Given Maclure's widely published ideas and admiration for the Society of Friends, it is hard to believe that Hicks was not cognizant of the New Harmony movement.[7] Taking his starting point from Isaiah, Hicks used animals to represent the botanical secrets of health, medicine, and life without the perturbations of strife. Painted from 1820 until Hicks's death in 1849, the many versions of the *Peaceable Kingdom* show a child or children standing among an unlikely cluster of animals. Often William Penn is present with Indians and other members of the Society of Friends in unexpected although distinctly American geological settings.

The Quaker aversion to worldly expressions, which included such pomp as the academic arts, suggests that these decorative paintings were primarily didactic.[8] According to this argument, the *Peaceable Kingdom* series repeats for dissemination Hicks's eclectic vision of human actions in the New World. Yet his idiom is so idiosyncratic that it has tended to defy placement within the American intellectual developments of his time.[9] Far from being anti-intellectual, all known sources for Hicks's imagery reflect the burgeoning market for elaborately engraved books, particularly Bibles and atlases, which Wilson's publication had inaugurated and Maclure's wealth perpetuated.

There are several good reasons to believe that some time after 1819 or 1820, Hicks visited the Philadelphia Museum—the most influential American institution for the promotion of natural histo-

Edward Hicks (1780–1849), ca. 1838. A reclining feline, lion, large ox, and dark bear are partly visible in the unfinished landscape of a *Peaceable Kingdom* in this portrait of the Quaker folk artist attributed to his nephew Thomas Hicks. (Abby Aldrich Rockefeller Folk Art Center)

ry at the popular level. Peale's Philadelphia Museum was by virtue of its uniqueness the logical place for a painter to go for visual information about animals, particularly the North American species Hicks added to his later pictures.[10] The museum's intellectual climate was compatible with Hicks's religious sentiments as well, for the main thrust of Peale's collections was educational rather than aesthetic. Furthermore, Peale admired the Society of Friends of which his last wife, Titian's stepmother, was a member, and he deliberately designed some exhibits to promote pacifistic relations with the Indians.

Hicks was an avid reader, and specific animals in his paintings can be traced to publications based on Peale's zoological exhibits. For example, a snoozing lion, a recurrent image in Hicks's pictures, also appears in a museum painting by Peale, and the museum may well have contributed to the fundamental concept of the *Peaceable Kingdom* series—the association of children with harmless stuffed beasts. Peale's museum and the natural history publications it fostered certainly had the potential to prepare viewers to accept Hicks's innovative animal iconography. In addition, the geological theories of the period clarify some of Hicks's oddly organized compositions, particularly those that introduce the Natural Bridge of Virginia or the Delaware Water Gap. Contemporary ideas regarding good health or peace of mind also shed considerable light upon Hicks's intense preoccupation with paintings, an activity that might well have subjected him to reproach within the Society of Friends. In contrast, Quaker interest in organized scientific activity was in no way unusual.[11] Penn, one of Hicks's heroes, had stressed the importance of training in natural history, and Hicks may have received elementary education in botany since George Fox urged that all Quaker schools provide instruction about "the nature of herbs, roots, plants and trees."[12]

In 1819, Hicks imitated the most famous naturalist of his day, Alexander Wilson, and undertook a trek to Niagara Falls. Seven years later, his souvenir of the trip, *The Falls of Niagara,* painted for Dr. Joseph Parrish, detailed the elemental concerns of American natural history. The extraordinary sight of Niagara Falls has remained a popular American tourist stop since Louis Hennepin's first description in 1679, and during the nineteenth century a number of artists, including John Vanderlyn and George Catlin, traveled to Niagara. Indeed, until recently, an engraving of Vanderlyn's painting of the falls was thought to be the model for Hicks's picture.[13] To the modern eye, the frame of Hicks's *Falls* is almost as striking as the picture for it contains boldly painted verses from Wilson's poem "The Foresters":[14]

> With uproar hideous first the Falls appear,
> The stunning tumult thundering on the ear.
>
> Above, below, where'er the astonished eye
> Turns to behold, new opening wonders lie,
>
> This great o'erwhelming work of awful Time
> In all its dread magnificence sublime,
>
> Rises on our view, amid a crashing roar
> That bids us kneel, and Times great God adore.

When Wilson composed these lines, he had not yet received the accolade his great volumes of ornithology would bring him, and he was by his own account suffering from severe bouts of depression.[15] His journey to the falls served as a spiritual diversion and was the first of a series of trips which he publicized by sending long descriptive narratives, often in the form of poems, to popular periodicals. The Philadelphia *Port Folio* eventually issued "The Foresters" in its entirety, which with one of Wilson's drawings established his reputation with Romantic readers as a wandering "Moral Philosopher."[16]

Ordinary readers found Wilson's approach to bird studies attractive because he combined scientific discovery with his interest in religion and literature.[17] As a result, his ornithological descriptions were widely quoted in the favorable press that accompanied the publication of each of his nine volumes from 1808 to 1814. Edward Hicks liked to read and write poetry as his sermons and picture frames show.[18] If he did not happen upon Wilson's verses in the *Port Folio* or in the early notices, he certainly was familiar with "The Foresters" after the poem's special printing in Newtown, where Hicks resided.[19] Having by this time failed at farming, Hicks embarked upon his inspirational quest and traveled to Niagara Falls the next year.

After his return from New York State, Hicks tried his hand at picture composition. Until this time, Hicks had confined his efforts to painting coaches and sign boards, but in 1820 he finished his earliest known *Peaceable Kingdom*.[20] Compelling visual evidence suggests that Hicks visited Peale's Philadelphia Museum before he completed this picture. The visual source of the recumbent leopard shown in this 1820 *Kingdom* (and repeated in many later versions) seems to be the leopard in a copy of Charles Catton's painting of *Noah's Ark,* which Charles Willson Peale exhibited in 1819.[21] Like Peale (and presumably Catton), Hicks has shown only the head and front half of the big cat stretched out sleepily in the lower right-hand corner of the picture. Although it is unlikely that Hicks saw Catton's original canvas (now lost), he may have been familiar with other works by the much-admired British painter. Elias Hicks (1748–1830), Edward's second cousin, owned a Catton "landscape with cattle and a figure," which was exhibited in New York City at the American Academy of Fine Arts in 1819.[22] That the lender, Elias Hicks, was perhaps the most famous Quaker preacher of his generation suggests that the Quaker proscription of fine art was not as rigid as some sources indicate. Edward stayed with the older Elias on Long Island in 1820, at a time when almost every study of Edward Hicks documents that he was searching for strong role models. Exposure to Catton's work in New York, as well as a desire to learn more about Wilson or landscape painting in general, may well have prompted Hicks to visit Peale's museum on his return to Philadelphia.

Many of Wilson's plates include finished landscapes. The background of plate 36, the bald eagle, for example, shows Niagara Falls and a rainbow engraved by Alexander Lawson for the fourth volume of the *American Ornithology* (1811).[23] In 1822, Peale painted the same eagle in the upper left-hand case of his *Portrait of the Artist in his Museum,* and Wilson's background of Niagara Falls was reproduced independently at least twice without the eagle.[24] A handsome engraving of the falls with a rainbow by George Cooke (1793–1849) credits the existence of another work by John J. Barralet (1747–1815) after a sketch by Wilson.[25] A would-be student of C. W. Peale, the precocious Cooke had engraved Wilson's *Port Folio* drawing of 1809, and Barralet, Lawson's former partner, drew a widely used profile of Wilson.[26] Unfortunately, the exact details of either man's connection with Wilson or the engraving for the *American Ornithology* will probably remain forever obscure because Ord destroyed many of Wilson's papers after his death.[27]

During the interval between 1820 and 1825, there was much besides Cooke's engraving to encourage Hicks to undertake his own Niagara picture. A fellow sign-painter in Philadelphia, Alvan Fisher (1792–1863), finished at least two paintings of the falls which are extremely similar to Cooke's engraving and even include Wilson's rainbow.[28] Equally important, in 1824 Ord's fine hand-colored edition of Wilson's *American Ornithology* reproduced the bald eagle plate and the poet's famous description. Wilson's volumes were far beyond the means of Hicks, who valued his own artistic labors at twenty-five cents a day.[29] But there were as many as seventy copies of the first edition alone in scientific libraries and prosperous homes in the Philadelphia area. Since Ord's edition generated renewed interest in Wilson's life and work, Hicks could easily have read the accompanying press notices at the Newtown Library Company, for which he made a sign in 1825, the same year he dated his *Falls of Niagara*.[30] On the basis of Wilson's influence alone, it would be reasonable to propose that Hicks copied the rattlesnake prominent in the lower left of his *Falls* from Wilson's 1809 *Port Folio* drawing; that Wilson's description of the eagle soaring above the falls in "The Foresters" led him to the Niagara plate in the *American Ornithology;* and that Hicks added the bald eagle as part of his tribute to Wilson so clearly expressed in the frame. Indeed, Hicks or his associates may have met Wilson when he approached the Quaker Friends of Wrightstown, New Jersey, about a teaching position.[31]

James E. Ayres has demonstrated that Hicks closely copied his *Falls of Niagara* from a different source—an 1822 map of the United States by Henry Schenck Tanner (1786–1858), which shows Niagara Falls and the Natural Bridge of Virginia.[32] Unlike Hicks's often whimsical adaptations of other artists' work, his *Falls* is an extremely precise enlargement of the right side of Tanner's vignette and reveals a degree of accuracy rare in Hicks's copied compositions. Following the War of 1812, maps, plates, and even scenic wallpapers were popular with travelers and stay-at-homes alike.[33] Foreign tourists made much of American geographical features, and by 1817 Maclure had prepared an enlarged second edition of his map of the United States.[34] In addition, the influx of such artists as John Hill and Charles Alexandre LeSueur from Europe gave additional impetus to mapmaking and other forms of illustration.[35] Philadelphia witnessed the greatest activity in this area because the city was the center for the engraving of bank notes, and in 1820 the engraving firm of Tanner, Vallance, Kearny & Company took on the publication of a large atlas of North America. This ambitious project, prompted in part by Wilson's monumental achievement, failed, but Tanner, the financial manager, took the plates and finished them to realize considerable personal profit at the expense of the other partners.[36]

Given Tanner's dubious scruples, we can only wonder about the originality of his Niagara design. His map, "constructed according to the latest information," certainly owed much to Maclure's published survey, and he may have had direct access to other scientific plates through engravers known to him. Tanner's partner, Francis Kearny (1785–1837), had worked with Peter Maverick in New York on the plates of the great quarto Bible, which provided the model for Hicks's first known *Peaceable Kingdom* of 1820. Both Maverick and Kearny later engraved John Godman's *American Natural History* and other museum-related publications.[37] Tanner's brother Benjamin (1775–1848) engraved for the *Port Folio* with Cooke and Barralet, and his teacher, Cornelius Tiebout, engraved natural history books printed at the museum and illustrated by Titian Peale in association with LeSueur.[38]

There was obviously a great deal of exchange and copying which occurred among illustrators at the level of the printer. At New Harmony, LeSueur designed a curtain for the village theater which showed the Falls of Niagara with a rainbow and rattlesnake. This design exists today only in the form of a theater ticket that was engraved at New Harmony by Tiebout sometime before his death in 1832.[39] To even the most casual eye, the overall similarity of this ticket to Hicks's *Falls of Niagara* is striking. The falls are shown from the same vantage point, and the treatment of such details as the rattlesnake and oversized bald eagle (here the left member of the pair) is close. Like Hicks's picture, this view of the falls even has an inscribed frame. Maclure donated a fine collection of books and plates to New Harmony, and given Maclure's geological interests, Tanner's plates might have been part of that gift. The map is not listed in that collection, however, and reference to it does not appear on the list of books removed to the Academy of Natural Sciences in Philadelphia after Maclure's death.[40] LeSueur's theater curtain was described by the observant Maximilian.[41] His artist, Karl Bodmer, appears to have been influenced by either Tanner's map or LeSueur's New Harmony design, for the elaborate map in Maximilian's North American travels book was published with a composite of images, the center of which is a falls with an oak tree, rattlesnake, beaver, and eagle above.[42]

Are the similarities of waterfalls by artists as disparate in training, ability, and purpose as Tanner, Hicks, LeSueur, Tiebout, and Bodmer coincidental or do they indicate a missing link? Tiebout and LeSueur may have seen Tanner's published map before their departure for New Harmony. Tiebout may also have had earlier access to Henry's vignette through his brother Benjamin. On the other hand, Henry Tanner certainly gained important ideas about book publication from Tiebout's example, and he may have copied an early Niagara design from him as well.[43] Again through Benjamin, Henry Tanner may have seen Wilson's unpublished drawings for the *Port Folio*. In any event, Wilson's rainbow, prominent in Cooke's engraving, persists in the LeSueur-Tiebout ticket designed for New Harmony. Of course, the possibility remains that their design was based on Hicks's painting. Tiebout and LeSueur gave local art lessons in the Philadelphia environs in which Hicks worked and may have shared patrons and engravers. Kearny engraved some of LeSueur's drawings, and his own drawing of the Quaker Richard Jordan of New Jersey resembles Hicks's earlier figure of William Penn.[44] Furthermore, Tiebout shared Hicks's admiration of landscape, and through Kearny or independently, he may have seen Hicks's *Falls of Niagara* before he moved to New Harmony.

The major elements of Hicks's *Peaceable Kingdom*s first appear in the *Falls*. An unnatural cluster of animals stands quietly in the foreground with a Quaker and Indians in the background. What the *Falls of Niagara* lacks is a historical component or covenant to unite these beings of creation in a more meaningful way than the Deism of Wilson's poem, and eventually Hicks used Penn's Treaty with the Indians as a straightforward pictorial device to associate Old Testament prophecy with New World reality.[45] Whereas the awesome cascades of Niagara had appealed to Wilson's romantic imagination, Hicks's interest was probably far more somber. Many early nineteenth-century writers viewed Niagara as the impressive vestige of an ancient geological cataclysm linked with Old Testament reprisals.[46] Hicks's *Peaceable Kingdom of the Branch* (now in the Yale University Art Gallery) also depicts a geological feature with religious significance. Like the *Falls,* this early picture belongs to the period 1825 to 1830, and like its predecessor, it uses a popular American tourist spot shown by Tanner, the Natural Bridge of Virginia. Here the verses around the frame, taken from the eleventh book of Isaiah, present the basic theme for all the *Peaceable Kingdom*s attributed to Hicks:

> The wolf also shall dwell with the lamb,
> & the leopard shall lie down with the kid;
> & the young lion & the fatling together; &
> a little child shall lead them. [11.6]

The juxtaposition of geology and morality, already evident in the *Falls,* becomes more emphatic here and in those *Peaceable Kingdom*s that employ the Delaware Water Gap, for like the celebrated Natural Bridge, contemporary thinkers reasoned that the gap was caused by retreating floodwaters.[47]

Hicks's association of geological wonder (the Natural Bridge) with human action (Penn's Treaty) and divine promise (the child) also reiterates the parallelism to the deluge found throughout Isaiah in such details as the receding stream, the withdrawing gray clouds, and, of course, the branch and animals on high ground.

Hicks's odd choice of locations is masterful. Explanation of dramatic formations like the Natural Bridge had confounded Hicks's scientific contemporaries, who usually sought causes consistent with the Mosaic account.[48] Even the great mind of the American Enlightenment, Thomas Jefferson, frankly confessed that the Natural Bridge made him "dizzy," and he was unable to provide a nonviolent account for its origin.[49] Eschewing Jefferson's cataclysmic geology, one thoughtful critic proposed the geological gradualism becoming fashionable in British scientific circles:

Mr. Jefferson's hypothesis rested entirely upon the supposition, that some sudden and violent convulsion of nature, tore away one part of the hill from the other, and left the bridge remaining over the chasm . . . instead of its being the effect of a sudden convulsion or an extraordinary deviation from the ordinary laws of nature, it will be found to have been produced by the very slow operation of causes which have always, and must ever continue, to act in the same manner.[50]

The slow operating cause the writer failed to name was simply erosion. Hicks's plain and peaceful composition may give the feeling of gradualism and the ordinary laws of nature, but the strange placement of Penn under the bridge suggests other, strongly worded warnings in Isaiah that the majesty of the Lord will "shake terribly the earth" (2.21), "And a man shall be as an hiding place from the wind, and a covert from the tempest, as rivers of water in a dry place, as the shadow of a great rock in a weary land" (32.2). Like Jefferson, Hicks apparently believed he was witnessing a landscape reformed by sudden, violent processes of the past. Unlike Jefferson, Hicks found reason for this violence in human sinfulness and social ills.

From 1827 to 1830, the schism occurring within the Society of Friends profoundly affected Hicks's outlook. He wrote, "It is a time when the great fundamental standards of Quakerism will have to be raised or raized & rallyed round if we expect as a society to be preserved from righthand errors & lefthand errors." Over the next decade, Hicks became increasingly concerned that the aftermath of the schism (which had centered on Elias Hicks) jeopardized his own ministry: "I cannot forsee the end & design of the storm gathering round me . . . I wish to have everything settled connected with my private character so that it may remain untarnished if my publick one is destroyed or as Jesse expresses it *'my way closed up.'* "[51] He declared the goals of his austere self-examination: to remove every "*root* & fibre of Jellocy envy & hatred" and to excoriate "that Harlot Self which has been the fatle Rock on which too many popular preachers have split and finely ruined." The modern reader can only conclude that Hicks was deeply troubled. His portrait by a young cousin, Thomas Hicks, certainly reveals the splenic visage that the leading Philadelphia medical authority attributed to "the Morbid Effects of Envy, Malice, and Hatred": "The face in this case performs the vicarious office which has lately been ascribed to the spleen . . . I once thought that medicine had not a single remedy in all its stores, that could subdue or even palliate the diseases induced by the baneful passions that have been described, and that an antidote to them was to be found only in religion." The author went on to recommend as palliatives the "convivial society between persons who are hostile to each other" and "putting their envious, malicious, and revengeful thoughts upon paper." Animals, even, were suggested, for they "suspend the anguish of the mind of this disease by their innocence, ingenuity or sports."[52]

Hicks was a textbook case, and his pictures were marked by these years of religious turmoil and mental distress. The animals frown; the razed landscape of the middle ground is barren. Hicks recognized his state of health and identified the creatures of the *Peaceable Kingdom* with the medical humors: "May the melancholy be encouraged and the sanguine quieted; may the phlegmatic be tendered and the choloric humbled. . . . The wolf also shall dwell with the lamb, and the leopard shall lie down with the kid."[53] Seen in this light, the animal groups represent the checked and balanced humors of a restored constitution and peaceful state of mind. The large ox included in so many of Hicks's compositions is, of course, the symbol of the evangelist Luke, the blessed physician. Painting was an exercise in mental hygiene, and the united elements of the *Peaceable Kingdom* provided a symbol of renewed spiritual health for both the individual and the Society of Friends at large.

Hicks's preoccupation with the increasingly complex message of his *Peaceable Kingdom* series did not preclude certain innovations. Hicks no longer restricted his choice of animals to the species mentioned in the Old Testament, and he included distinctive North American fauna such as the short-tailed bobcat, the dark phase of the red fox, the grizzly bear, and possibly the prairie wolf or coyote. These New World forms, of course, have no biblical reference, and thus Hicks had to rely once again on a secular source—the only comprehensive American book illustrating the mammals of this country, Godman's popular *American Natural History* (1826–28).

Like Wilson, whose work he emulated, Godman's descriptions relied on specimens at Peale's museum. Of the exhibited mammals, the mounted grizzly bears attracted special attention because they were the only examples to be seen east of the Mississippi River. Many of Hicks's paintings simply show the head of a bear eating corn with a cow, but he also showed an adult bear standing with one forearm over a tree limb. In these pictures, the adult lacks the emblematic quality of her griffon-pawed cub, and Godman's first volume illustrates Hicks's probable model, a plate drawn by LeSueur and engraved by Kearny. Titian Peale had also drawn these same bears, which had been donated by Jefferson and mounted by his father. Hicks could have seen either Titian's drawing or Kearny's engraving. Both are similar, and both were available in the Philadelphia area. Larger editions of the *American Natural History* with plates were published after Godman's death in 1830, the same year that Titian's drawing was lithographed for the *Cabinet of Natural History and American Rural Sports*.[54] Like James Otto Lewis's magazine *Aboriginal Portfolio,* the now obscure *Cabinet* was very much a part of the local art scene. Published by the artist brothers John and Thomas Doughty, it included articles by or about John Neagle, George Catlin, and Titian Peale's friend Thomas Sully (whose 1819 painting of George Washington was copied by Hicks).[55]

The extent of Hicks's concern with the animals and Indians of the American wilderness should not be exaggerated. Godman's volumes, however, contained much information to pique the reader's interest. The first plate in Godman's book is Peter Maverick's engraving of Neagle's portrait of Petalasharoo, which is accompanied by a long discussion of the American Indians. Like Peale and Wilson, Godman was fascinated by the effects of white settlement upon animal behavior, and his text contained charming accounts of the amelioration of animal instincts. For example, Godman's description of a young fox nurtured by a common house cat at the museum was just the peaceful cohabitation Hicks hoped for in another state of nature.[56] Could Hicks have been influenced by this incident? The fox was engraved on the same plate as a sleepy prairie wolf, shown lying down. Hicks painted his wolves lying down, except in later versions in which he made a telling mistake. The requisite "wolf" standing gracefully in his *Kingdom*s of 1848 and 1849 is really the large dark fox illustrated on another plate in Godman's *American Natural History.*

The visual information available to the most casual reader of Godman's volumes was also on display to the public at the Peale museum. After 1820, Titian's drawings of western species provided the basis for new mounts, and some of his watercolor studies were exhibited at the museum and also at the Academy of Art.[57] The prairie wolf (coyote) shown in Godman's book had been captured by Titian on the upper Missouri and sent back live to the museum,[58] where it lived for a while in Peale's menagerie. Peale's Quadruped Room also offered possible models for the big cats shown in Hicks's paintings of 1830, 1833, and 1835, which resemble plates in Godman's book and Titian's earlier field study of the western bobcat. As for Old World species, exotic felines like the leopard which are prominent in all Hicks's *Peaceable Kingdom*s were on continuous display after 1797, and museum records state that Peale's dioramas included a "Wolf and lamb group."[59] The availability of such models in no way conflicts with the idea that Hicks first copied animals from a plate published in Bibles. What is curious is that, out of the sixty engravers employed in Philadelphia, the Bible plate Hicks used was engraved by P. Maverick, Durand & Co. and Tanner, Vallance, Kearny & Co.

The image of the child continued to figure importantly in every one of Hicks's *Peaceable Kingdom*s for almost thirty years, and like

Peale's exhibits and Maclure's educational programs, some of his pictures may have been intended for children.[60] Painting children with animals was a common and endearing device of itinerant limners during Hicks's lifetime.[61] Charles Willson Peale's formal contributions to this charming association cannot be overlooked. Peale encouraged parents to bring their children to the museum. His many portraits of children showed them with plants or animals, and he recounted instances of young members of his family with wild species ranging from the notorious grizzlies and carnivorous hyena to field mice.[62] He obviously enjoyed seeing children with animals. When Peale moved the museum collections in 1795, "he collected all the boys of the neighborhood" to publicize the new location, and the parade, with its "Panthers, Tyger Catts and a long string of Animals of smaller size carried by the boys," bore no small resemblance to Hicks's later *Noah's Ark* or his 1849 *Kingdom,* which shows a boy pulling a bobcat on a string.[63] Children were truly an integral part of the museum's image and outreach. Peale's wonderful staircase painting of his two young sons hung for many years at one end of the museum, and an early formal acknowledgment designed by Peale for museum donors shows a child raising the curtain of nature on an imaginary landscape of animals in much the same way as the artist painted himself raising the curtain upon his museum twenty years later.

Peale promoted his museum as a "Book of Nature" to be enjoyed by all. In the background of his famous self-portrait of 1822, a young man, perhaps a "scientific gentleman," stands pensively, a father instructs his son, and a woman marvels at the reconstructed mastodon. In an earlier portrait of his mother, Peale lovingly painted her showing a book of botanical drawings (which look like the work of William Bartram) to her grandchildren, Rembrandt, Raphael, and Angelica.[64] Peale's picture celebrates grandparenthood, that most notable invention of eighteenth-century public health, and Peale was himself a great advocate of posterity. He was so proud of the *American Ornithology* that he sent a copy to London to the greatest American artist alive, Benjamin West.[65] At a time when young women exhibited needlework patterns based on Wilson's plates in Philadelphia, it is difficult to believe that Hicks was not availing himself of the resources developed by Peale's museum and disseminated through Maclure's patronage of science, education, and the printing arts.[66]

Although the wild and exotic species in Hicks's pictures offer historians the greatest challenge, the large oxen characteristic of so many of his *Peaceable Kingdom*s cannot be overlooked. All the domestic animals in his *Noah's Ark, Washington at the Delaware,* farm scenes, and pastoral landscapes are massive and extremely robust. In the well-ordered barnyard of *The Residence of David Twining,* Hicks's childhood home, a Negro plows, the porkers eat in a row, and although a dog barks at a spitting cat, the fat horses stand quietly to be mounted. On this properly managed estate that Hicks recreated from memory, the animals are large, healthy, and fecund, and the people are upright, active, and prosperous. The architectural modifications are pleasing to the eye, and the fields are well drained and tended. Hicks has taken the viewer back to the year 1787, just six years after Jefferson had penned the *Notes on the State of Virginia.* Given his later interest in books, he has tellingly shown himself as a small boy reading beside the knee of his foster mother, and in this, one of his last pictures, Hicks has returned to the early raison d'être of the American natural history he so easily assimilated and so obviously admired. The Twining residence could serve equally well as an illustration for Jefferson's refutation of Buffon in query six of the *Notes* or as an advertisement for Owen and Maclure's New Harmony project. Like his other *Peaceable Kingdom*s, this painting bears a prophetic strain, for the Owen family developed prize-winning shorthorn cattle, and by the late nineteenth century, their model New Harmony farms came to resemble Hicks's pastoral ideal.

In another sense, Nuttall's personal objective for his profession, "natural desire for harmony," also found "accord" in Hicks's prolific industry. The painted compositions, however awkwardly executed, repeat the same overriding confidence in the North American landscape as a suitable nursery for the "uncontaminated gift of Nature." For Hicks, this "last and most perfect of systems" was divine manifestation. For Nuttall and his colleagues, it was the revelation of the natural system of classification, and for Maclure, of course, the "uncontaminated gift of Nature" was the human mind free to enjoy the benefits of science in a democratic society. Although the more gifted Peales may have viewed Hicks's pictorial efforts with artistic disdain, Hicks certainly would have responded to Titian's journal entry for 1819 of a wild goose in "peaceable possession of an eagle's nest." A boy (eighteen-year-old Peale) standing near the banks of the great Missouri River with an unlikely association of animals before him and Quakers conversing with Indians (Say and the Iowa tribes) behind him was exactly the juxtaposition of historic fact, geological setting, and zoological detail Hicks sought in his promulgation of the Peaceable Kingdom.

"Harmonie," wrote W. C. Peters in an ode for the Rappites, Maclure's predecessors at New Harmony, "citizen of a better world." "All creatures are satisfied," the young Englishman continued, and his words might well have served to caption Hicks's painting of 1832, which shows "Quakers bearing banners" behind a group of contented beasts.[67] While Maclure's schools developed citizens of "a better world" through science and Hicks used museum materials to describe a fourth peaceable kingdom of nature, other Philadelphians, some also from Quaker families, put aside the three kingdoms of Linnaeus and reorganized natural history to accommodate classification, religion, and conservative social mores.

Notes

Introduction: An Invitation

1. Baldwin Möllhausen, *Diary of a Journey from the Mississippi to the Coast of the Pacific with a United States Government Expedition,* trans. Mrs. Percy Sinnett, 2 vols. (London: Longman, Brown, Green, Longmans & Roberts, 1858), 1:3.
2. Marjorie H. Nicolson, *Mountain Gloom and Mountain Glory* (Ithaca: Cornell University Press, 1959), pp. 12–16.
3. John Conran, ed., *The American Landscape: A Critical Anthology of Prose and Poetry* (New York: Oxford University Press, 1974), pp. 141, 143.
4. Letter, December 13, 1820, quoted by John Francis McDermott, ed., *Audubon in the West* (Norman: University of Oklahoma Press, 1965), p. 7.
5. The quotations are from Samuel Taylor Coleridge, "Kubla Khan: or, A Vision in a Dream," line 12, "This Lime-Tree Bower My Prison," line 12, and William Wordsworth, "Lines Composed a Few Miles above Tintern Abbey," line 6, all in W. H. Auden and Norman H. Pearson, eds., *Romantic Poets: Blake to Poe* (New York: Viking Press, 1968), pp. 153, 160, 192.
6. Harry Weiss and Grace Ziegler, "The Communism of Thomas Say," *Journal of the New York Entomological Society* 35 (September 1927):231–39, is an interesting early interpretation of Say's loyalty to Maclure.
7. Francis Harper, ed., *The Travels of William Bartram: Naturalist's Edition* (New Haven: Yale University Press, 1958), p. 106. Bartram's title gives an accurate outline of the book's contents: *Travels through North and South Carolina, Georgia, East and West Florida, the Cherokee Country, the Extensive Territories of the Muscogulges, or Creek Confederacy, and the Country of the Choctaws: Containing an Account of the Soil and Natural Productions of These Regions, Together with Observations on the Manners* (Philadelphia: James and Johnson, printers, 1791). Bartram's *Travels* was promptly reprinted in England and Ireland and translated into German, Dutch, and French.

8. See William Bartram's "Anecdotes of an American Crow," *Philadelphia Medical and Physical Journal* 1 (1805):89–95.
9. Thomas Nuttall, *A Popular Handbook of the Ornithology of Eastern North America,* ed. Montague Chamberlain, 2d rev. ed., 2 vols. (Boston: Little, Brown, 1896), 1:xxxvi–vli, originally published with woodcuts under the title *A Manual of the Ornithology of the United States* in 1832, a rare book I have not seen.
10. William Cullen Bryant, "The Embargo, or Sketches of the Times; a Satire," printed in Boston in 1808 and quoted in Darrel Abel, ed., *American Literature: Colonial and Early National Writing* (Woodbury, N.Y.: Barron's Educational Series, 1962), p. 396.
11. Hugo A. Meier, "Thomas Jefferson and a Democratic Technology," in Carroll W. Pursell, Jr., ed., *Technology in America* (Cambridge, Mass.: MIT Press, 1982), p. 23. See also Brooke Hindle, *The Pursuit of Science in Revolutionary America, 1735–1789* (1956; rpt. New York: W. W. Norton, 1974), pp. 131, 133.
12. Charles Coleman Sellers, *Mr. Peale's Museum* (New York: W. W. Norton, 1980), p. 101.
13. Edward T. Martin, *Thomas Jefferson: Scientist* (New York: Henry Schuman, 1952), p. 118, discusses Mrs. Royall's visit.
14. Thomas Jefferson, *Notes on the State of Virginia* (Philadelphia: R. T. Rawle, 1801), p. 42, presumably referring to Georges Louis LeClerc, comte de Buffon, *Histoire naturelle, générale et particulière,* 15 vols. (Paris: L'imprimerie royale, 1749–67), 9:109–10, first published in 1761. See also D. Boehm and E. Schwartz, "Jefferson and the Theory of Degeneration," *American Quarterly* 9 (1957):454–59.
15. See Benjamin Smith Barton to William Bartram, November 30, 1803, cited by Nathan B. Fagin, *William Bartram: Interpreter of the American Landscape* (Baltimore: Johns Hopkins University Press, 1933), p. 12.
16. A. O. Weese, ed., "The Journal of Titian Ramsay Peale, Pioneer Naturalist," *Missouri Historical Review* 41 (1947):284, entry for April 26, 1820, near Council Bluffs.
17. Titian R. Peale to R. M. Patterson, November 13, 1838, written at sea, Coleman Sellers Mills Collection, American Philosophical Society, Philadelphia, in Jessie Poesch, *Titian Ramsay Peale and His Journals of the Wilkes Expedition* (Philadelphia: American Philosophical Society, 1961), p. 72; see also William Stanton, *The Great United States Exploring Expedition of 1838–1842* (Berkeley and Los Angeles: University of California Press, 1975), pp. 29, 75–76, 84–85.
18. Amos Eaton to John Torrey, January 12, 1822, quoted by Ethel M. McAllister, *Amos Eaton, Scientist and Educator, 1776–1842* (Philadelphia: University of Pennsylvania Press, 1941), p. 218. For another view, see William Smallwood, "Amos Eaton, Naturalist," *New York History* 18 (1937):167–88.

Chapter 1: The Quality of Nature

1. Harcourt Brown, "Buffon and the Royal Society of London," in M. F. Ashley Montagu, ed., *Studies in the History of Science and Learning* (1944; rpr. New

York: Krause, 1969), p. 145. Otis E. Fellows and Stephen F. Milliken, *Buffon* (New York: Twayne, 1972), pp. 15–16. Early English citations include the widely used interpretation of Buffon's work by William Smellie, *Natural History, General and Particular,* 9 vols. (London: W. Strahan and T. Caddell; Edinburgh: W. Creech, 1770; 2d ed., 1785), and also the entire *Natural History of Animals, Vegetables and Minerals with the Theory of the Earth in General,* 6 vols. (London: T. Bell, 1775).

2. John Lyon and Phillip R. Sloan, *From Natural History to a History of Nature* (Notre Dame: University of Notre Dame Press, 1981), pp. 1–9.

3. In 1765, Garden wished that Buffon's ideas agreed with the Linnaean method; see his letter to John Ellis in James Edward Smith, ed., *A Selection of the Correspondence of Linnaeus and Other Naturalists,* 2 vols. (London: Longman, Hurst, Rees, Orme, and Brown, 1821), 1:546.

4. Durand Echeverria, *Mirage in the West* (New York: Octagon Books, 1966), pp. 9–14.

5. Benjamin Smith Barton, "An Inquiry into the Question, Whether the Apis Mellifica or True Honey-Bee Is a Native of America," *Transactions of the American Philosophical Society* 3 (1793):242, 250.

6. Unless otherwise noted, references are to the early American edition of Thomas Jefferson, *Notes on the State of Virginia* (Philadelphia: R. T. Rawle, 1801), pp. 49–141; see Coolie Verner, *A Further Checklist on the Separate Editions of Jefferson's Notes on the State of Virginia* (Charlottesville: Bibliographical Society of the University of Virginia, 1950), pp. 13–21.

7. In the *Notes,* pp. 88–89, Jefferson directs the reader to Georges Louis Leclerc, comte de Buffon, *Histoire naturelle général et particulière,* 15 vols. (Paris: L'imprimerie royale, 1749–67), 9:87, 102, 105. For aspects of Buffon's theory not treated here, see Paul L. Farber, "Buffon and the Concept of Species," *Journal of the History of Biology* 5 (1972):259–84; Peter J. Bowler, "Bonnet and Buffon: Theories of Generation and the Problem of Species," *Journal of the History of Biology* 6 (1973):259–81; and J. S. Wilkie, "The Idea of Evolution in the Writings of Buffon, Part II," *Annals of Science* 12 (1956):212–27.

8. Jefferson, *Notes,* p. 87.

9. For the extensive literature on Jefferson's paleontology, see Roland W. Brown, "Jefferson's Contributions to Paleontology," *Journal of the Washington Academy of Science* 83 (1943):257–59; Henry Osborn, "Thomas Jefferson as a Paleontologist," *Science* 82 (1935):533–37; the classic George Perkins Merrill, *The First One Hundred Years of American Geology* (New Haven: Yale University Press, 1924), p. 16, is the main source for E. T. Martin, *Thomas Jefferson: Scientist* (New York: Henry Schuman, 1952), pp. 107–17; and George Gaylord Simpson, "The Beginnings of Vertebrate Paleontology in North America," *Proceedings of the American Philosophical Society* 86 (1942):130–57.

10. Jefferson, *Notes,* p. 79. See also his "Memoir on the Discovery of Certain Bones of a Quadruped of the Clawed Kind in the Western Parts of Virginia," *Proceedings of the American Philosophical Society* 4 (1799):256, 246–60, an explanation revived by Paul S. Martin, "The Discovery of America," *Science* 179 (March 1973):969–74; James E. Mosimann and Paul S. Martin, "Simulating Overkill by Paleo Indians," *American Scientist* 63 (May–June 1975): 304–13.

11. Samuel L. Mitchill, *Discourse on Thomas Jefferson More Especially as a Promoter of Natural and Physical Sciences* (New York: G. C. Carvill, 1826), p. 22.

12. See Jefferson, *Notes,* pp. 95–96, 97–112, 122, 141.

13. Gilbert Chinard, "The American Philosophical Society and the Early History of Forestry in America," *Proceedings of the American Philosophical Society* 89 (1945):444–87; see also Chinard's "Eighteenth Century Theories of America as a Human Habitat," *Proceedings of the American Philosophical Society* 91 (1947):27–57; William J. Humphreys, "A Review of Papers in Meteorology and Climatology Published by the American Philosophical Society prior to the Twentieth Century," *Proceedings of the American Philosophical Society* 86 (1942):29–33.

14. William Bartram merits his own biography; see Robert McCracken Peck, "William Bartram and His Travels," in Alwyne Wheeler, ed., *Contributions to the History of North American Natural History* (London: Society for the Bibliography of Natural History, 1983), p. 38.

15. Francis Harper, ed., *The Travels of William Bartram: Naturalist's Edition* (New Haven: Yale University Press, 1958), pp. 104, 121.

16. Ibid., p. lvi.

17. Ibid., p. lxi.

18. William Barton, "Observations on the Probabilities of the Duration of Human Life and the Progress of Populations, in the United States of America," *Transactions of the American Philosophical Society* 3 (1793):25–61, note charts, pp. 55–61, and botanical postscript, pp. 134–38. He cites Benjamin Franklin, *Observations Concerning the Increase of Mankind and the Peopling of Countries* (Boston: Kneeland, 1755).

19. See Meyer Reinhold, "The Quest for 'Useful Knowledge' in Eighteenth-Century America," *Proceedings of the American Philosophical Society* 119 (April 1975):108–32.

20. Alexis de Tocqueville, *Democracy in America,* trans. George Lawrence (New York: Doubleday, 1969), p. 459, a passage discussed by Hugo Meier, "Thomas Jefferson and a Democratic Technology," in Carroll Pursell, Jr., ed., *Technology in America* (Cambridge, Mass.: MIT Press, 1982), p. 23; see also Nathan Reingold, "Reflections on 200 Years of Science in the United States," *Nature* 262 (July 1976):9.

21. Peter Gay, *The Party of Humanity* (New York: W. W. Norton, 1959), pp. 119–20ff. For American reaction to the beast machine, see William Bartram, "Anecdotes of an American Crow," *Philadelphia Medical and Physical Journal* 1, pt. 1. (1805):89–95.

22. Gareth Nelson, "From Candolle to Croizat: Comments on the History of Biogeography," *Journal of the History of Biology* 11 (Fall 1978):273–78. For application of Buffon's "northern principle" to American species, see J. W. [James Wilson?] "Review," manuscript dated October 5, 1830, from the library of Sir William Jardine, Alexander Wilson Papers, Houghton Library, Harvard University; see also Charles Lucien Bonaparte, *American Ornithology, or the Natural History of Birds Inhabiting the United States Not Given by Wilson,* 4 vols. (Philadelphia: S. A. Mitchell; Carey, Lea, & Carey, 1825–32), 2:20, 40; Constantine Rafinesque, *New Flora and Botany of North America* (1836; rpt. Cambridge, Mass.: Murray, 1946), pp. 30–31; and Leonard Huxley, ed., *Life and Letters of Joseph Dalton Hooker* (London: J. Murray, 1918), p.224.

23. René Taton, ed., *The Beginnings of Modern Science from 1450 to 1800,* trans. A. J. Pomerans, 2 vols. (New York: Basic Books, 1966), 2:620–22.

24. See Coolie Verner, *Mr. Jefferson Distributes his Notes* (New York: New York Public Library, 1932), pp. 9–25.

25. François André Michaux, *The North American Sylva,* trans. Augustus L. Hillhouse, 3 vols. (Paris: C. D. Hautel for T. Dobson and G. Conrad of Philadelphia, 1818, 1819).

26. See Reuben Gold Thwaites, "Preface," to André Michaux, *Journal of André Michaux,* in Thwaites, ed., *Early Western Travels,* vol. 3 (Cleveland: Arthur H. Clark, 1904), pp. 11–15. The American Philosophical Society received the original manuscript after Michaux's death.

27. François André Michaux, *The North American Sylva,* ed. J. J. Smith, vol. 1 (Philadelphia: Rice, Rutter, 1865), pp. 13, 15, 25, 27, 29.

28. Ibid., p. 28.

29. Ibid., pp. 37, 70, 57, 23, 28 (see also pp. 51, 54–55), 66 (see also p. 108).

30. See F. André Michaux to William Bartram, March 12, 1810, in William Darlington, *Memorials of John Bartram and Humphrey Marshall* (1849; facsimile ed., New York: Hafner, 1967), p. 477; Alexander Wilson to William Bartram, October 25, 1809, Wilson Scrapbook, Houghton Library; Alexander Wilson, *American Ornithology, or the Natural History of Birds of the United States,* 9 vols. (Philadelphia: Bradford and Inskeep, 1808–14), 3:viii; Samuel George Morton, *A Memoir of William Maclure, Esq.* (Philadelphia: T. K. and P. G. Collins, 1841), p. 31.

31. DeWitt Clinton, *An Inventory Discourse Delivered before the Literary and Philosophical Society of New-York on the Fourth of May, 1814* (New York: David Longworth, 1815), p. 2.

32. Hibernicus, *Letters on the Natural History and Internal Resources of the State of New York* (New York: Sold by E. Bliss & White, 1822), pp. 39, 11.

33. Hans Huth, *Nature and the American* (Lincoln: University of Nebraska Press, 1972), p. 32.

34. Charles Willson Peale, *Scientific and Descriptive Catalogue of Peale's Museum* (Philadelphia: Samuel H. Smith, 1796), p. viii.

35. See George P. Merrill, "Contributions to a History of American State Geological and Natural History Surveys," *Bulletin of the U.S. National Museum,* no. 109 (1920), pp. 327–41.

Chapter 2: The Lessons of Nature

1. Comprehensive studies are Charles Coleman Sellers, *Charles Willson Peale,* 2 vols. (Philadelphia: American Philosophical Society, 1947), and his *Mr. Peale's Museum* (New York: W. W. Norton, 1980); see also Edgar P. Richardson, Brooke Hindle, and Lillian B. Miller, *Charles Willson Peale and His World* (New York: Harry N. Abrams, 1982).

2. Paul Lawrence Farber, "The Development of Taxidermy and the History of Ornithology," *Isis* 68 (1977):552–53.

3. See Titian Ramsay Peale's untitled and unsigned introduction to the *Cabinet of Natural History and American Rural Sports* 1 (1830):vi.

4. For an inventory of extant specimens see Walter Faxon, "Relics of Peale's Museum," *Bulletin of the Museum of Comparative Zoology* 59 (July 1915):119–

33. For notes on Peale's bird's eggs at Harvard University, see Elsa Guerdon Allen, "The History of American Ornithology before Audubon," *Transactions of the American Philosophical Society,* n.s., 41 (1931):569.

5. James Flexner, *The Light of Distant Skies* (New York: Dover, 1969), pp. 97–107.

6. [Peale] *Cabinet,* p. vi.

7. Charles Willson Peale, *Discourse Introductory to a Course of Lectures on the Science of Nature with Original Music* (Philadelphia: Zacariah Poulson, 1800), p. 3.

8. Quoted in John C. Greene, "The Founding of Peale's Museum," in Thomas R. Buckman, ed., *Bibliography and Natural History* (Lawrence: University of Kansas Libraries, 1966), p. 67.

9. Wilson's *American Ornithology, or the Natural History of the Birds of the United States* was published in nine volumes between 1808 and 1814 by Bradford and Inskeep and printed by R. and W. Carr of Philadelphia. See vol. 2 (1810), pp. vi–vii.

10. See statements following the Minutes of 1812, Academy of Natural Sciences Archives, cited by Patsy Gerstner, "The Academy of Natural Sciences in Philadelphia," in Alexandra Oleson and Sanborn C. Brown, eds., *The Pursuit of Knowledge in the Early American Republic* (Baltimore: Johns Hopkins University Press, 1976), p. 176.

11. Ralph S. Bates, *Scientific Societies in the United States* (Cambridge, Mass.: MIT Press, 1965), p. 43; Maurice E. Phillips, "The Academy of Natural Sciences of Philadelphia," *Transactions of the American Philosophical Society,* n.s., 43 (1953):266–71. Also see Samuel G. Morton, "History of the Academy of Natural Sciences of Philadelphia," *American Quarterly Register* 13 (1841):433–38.

12. Raymond Phineas Stearns, *Science in the British Colonies of America* (Urbana: University of Illinois Press, 1970), pp. 506–7. See also Ann Leighton, *American Gardens in the Eighteenth Century* (Boston: Houghton Mifflin, 1976), pp. 75–101, and Blanche Henrey, *British Botanical and Horticultural Literature before 1800,* 3 vols. (London: Oxford University Press, 1975).

13. See Charles Willson Peale's delightful *Scientific and Descriptive Catalogue of Peale's Museum* (Philadelphia: Samuel H. Smith, 1796).

14. For more details see Tore Frangsmayr, ed., *Linnaeus: The Man and His Work* (Berkeley and Los Angeles: University of California Press, 1983), pp. 1–109.

15. Translated by W. C. Turton (London, 1806) and included in Thomas Hall, ed., *A Source Book in Animal Biology* (Cambridge, Mass.: Harvard University Press, 1970), p. 32; see also Sven Horstadius, "Linnaeus, Animals and Man," *Biological Journal of the Linnean Society* 6 (December 1974):269–75; and Erik Nordenskiöld, *The History of Biology* (New York: Tudor, 1935), pp. 203–17.

16. Frank Davison Adams, *The Birth and Development of the Geological Sciences* (New York: Dover, 1954), pp. 200–201, 220–26, 239–49; see also Sir Gavin DeBeer, "Biology before the Beagle," in Philip Appleman, ed., *Darwin* (New York: W. W. Norton, 1970), pp. 3–10.

17. W. N. Blane, *Travels through the United States and Canada* (London: Baldwin, 1828), p. 22–23.

18. A Hunter Dupree, *Science in the Federal Government* (Cambridge, Mass.: Belknap Press of Harvard University Press, 1957), pp. 27–28.

19. Gerstner, "The Academy," pp. 174–77.
20. Quoted in Julia Lewis Morris, *From Seed to Flower* (Philadelphia: Pennsylvania Horticultural Society, 1976), p. 61, from material at the Hunt Botanical Library, Pittsburgh.
21. Marilyn S. Milcher, "Round Panel Furniture of Virginia's Eastern Shore, 1730–1830," *Art & Antiques* 5 (November–December 1982):85.
22. Sellers, *Museum,* p. 101.
23. See the unsigned article in *The Cabinet of Natural History and American Rural Sports* 1 (1830):1–2.
24. Given the attention to invertebrates by American naturalists, especially those working in the South, the early influence of Lamarck in the United States merits scholarly study.
25. Hermann Neal Fairchild, *History of the New York Academy of Sciences* (New York: Published by the author, 1887), p. 11.
26. William Dandridge Peck, "Four Remarkable Fishes Taken Near the Piscotaqua, N.H.," *Memoirs of the American Academy of Arts and Sciences* 2 (1797):46.
27. Benjamin Henry Latrobe, "A Drawing and Description of the Clupea Tyrranus and Oniscus Praegustator," *Transactions of the American Philosophical Society* 5 (1802):77–81, dated December 18, 1799.
28. See Samuel Latham Mitchill, *Report in Part on the Fishes of New York,* ed. Theodore Gill (Washington, D.C.: Printed for the editor, 1898), originally published January 1, 1814, and continued in the *Transactions of the Literary and Philosophical Society of New York* 1 (1815); see also Mitchill, *The Picture of New York, or The Traveller's Guide through the Commercial Metropolis of the United States* (New York: Riley, 1807), pp. 130–34. Mitchill erroneously used Latrobe's suggested binomial for the edible alewife for the inedible menhaden.
29. See, for example, C. Rafinesque Schmaltz, "Progress in American Botany," *Medical Repository* 1 (1810):297. The next year Mitchill published Rafinesque's "Essay" on exotic plants, *Medical Repository* 2 (1811):330–45.
30. C. S. Rafinesque, *Ichthyologia Ohiensis* (Lexington: Printed for the author by W. G. Hunt, 1820), p. 7, and "The Fishes of the United States," *Atlantic Journal and Friend of Knowledge* 1 (1832):142–43.
31. Constantine Rafinesque, "Fragments of a Letter to Mr. Bory St. Vincent at Paris," *Western Minerva* 1 (1821):70–71. This journal exists only as a new printing pulled from the old plates at the Academy of Natural Sciences.
32. Louis Agassiz, "Synopsis of the Ichthyological Fauna of the Pacific Slope of America," *American Journal of Science,* n.s., 19 (1855):91. For an example of a more recent assessment, see Milton B. Trautman, *The Fishes of Ohio* (Baltimore: Waverly Press, 1957), p. 156.
33. See George Ord, "A Memoir of Charles-Alexandre LeSueur," *American Journal of Science,* 3d ser., 8 (September 1849):189–216.
34. See Say's Appendix to chapter 16 of Edwin James, *Account of an Expedition from Pittsburgh to the Rocky Mountains Performed in the Years 1819, 1820,* 2 vols. (Philadelphia: H. C. Carey and I. Lea, 1823), 1:369ff., and Say's field notebook at the Museum of Comparative Zoology, Harvard University, Cambridge, Massachusetts.
35. William Keating, *Narrative of an Expedition to the Source of St. Peter's River, Lake*

Winnepeck, Lake of the Woods, &c, &c, (Philadelphia: H. C. Carey and I. Lea, 1824), pp. 254–55.

36. See Francis Harper, ed., *The Travels of William Bartram: Naturalist's Edition* (New Haven: Yale University Press, 1958), pp. 81, 167.

37. Benjamin Smith Barton, "A Memoir Concerning the Fascinating Faculty Which Has Been Ascribed to the Rattlesnake and Other American Serpents," *Transactions of the American Philosophical Society* 4 (1799):74–113. Latrobe's drawing was discovered at the American Philosophical Society by the staff of the Papers of Benjamin Henry Latrobe, Edward C. Carter, editor.

38. George Ord to Charles Waterton, April 23, 1832, microfilm, American Philosophical Society.

39. H. H. Porter, "Preview of the Ornithological Biography," *Monthly American Journal of Geology* 1 (September 1831):136–39. For the pros and cons, see Col. Albert to Richard Harlan, letter extract, cited by Richard Harlan, "Description of *Vespertilio Auduboni,* a New Species of Bat," ibid. (November 1831):221–23.

40. Edwin James, *Account of an Expedition from Pittsburgh to the Rocky Mountains Performed in the Years 1819, 1820,* 3 vols. (London: Hurst, Rees, Orme & Brown, 1823), 1:45–47, 3:43n. See also James, *Account,* ed. Reuben Gold Thwaites, *Early Western Travels,* vol. 15 (Cleveland: Arthur H. Clark, 1905), pp. 324–25, for valuable commentary.

41. Dupree, *Science in the Federal Government,* pp. 27–28.

42. Donald Jackson, ed., *Letters of the Lewis and Clark Expedition with Related Documents* (Urbana: University of Illinois Press, 1962), pp. 267–68.

43. For one account, see Alexander Wilson, "The Particulars of the Death of Captain Lewis," *Port Folio* 7 (1812):34–47.

44. For example, the Lewis and Clark zoological specimens are not even discussed in the otherwise excellent Francis W. Pennell, "Benjamin Smith Barton as a Naturalist," *Proceedings of the American Philosophical Society* 86 (September 1942):108–22.

45. For the best discussion see Jeannette E. Graustein, *Thomas Nuttall, Naturalist* (Cambridge, Mass.: Harvard University Press, 1967), pp. 32–41, 85–86, 103–4.

46. George Ord, "North American Zoology," in William Guthrie, ed., *A New Geographical, Historical and Commercial Grammar* (Philadelphia: Johnson & Walker, 1815), pp. 291–313. Compare also Paul Russell Cutright, *Lewis and Clark: Pioneering Naturalists* (Urbana: University of Illinois Press, 1969), pp. 438–47, and Jackson, ed., *Letters,* pp. 293–97.

47. In 1815, George Ord carefully designated the red-breasted squirrel *Sciurus rubicatus,* and his nomenclature for the prairie dog and grizzly bear also gained common usage. In the Paris *Journal de Physique* 88 (1818):146–55, he named the Columbian gray squirrel *S. griseus* and the Rocky Mountain ground squirrel *S. trocloclytus.* He realized that the so-called brown squirrel was the Hudson Bay squirrel and described the ash-colored rat (*Mus cinerus*) in the *Journal of the Academy of Natural Science* 5 (1825):346.

48. *Arvicola Nuttall,* a gerbil-like creature, described by Richard Harlan in the *Monthly American Journal of Geology* 1 (April 1832):446–47.

49. See Jeannette E. Graustein, "Audubon and Nuttall," *Scientific Monthly* 74 (February 1952):86.

50. Charles Willson Peale, *Scientific and Descriptive Catalogue,* p. 28.
51. Jessie Poesch, *Titian Ramsay Peale and His Journals of the Wilkes Expedition* (Philadelphia: American Philosophical Society, 1961), p. 48.
52. Titian Peale's preparatory drawing for the mule deer is among the Titian Ramsay Peale manuscripts at the American Philosophical Society, Philadelphia; Thomas Nuttall, *A Manual of the Ornithology of the United States and Canada,* vol. 1 (Cambridge, Mass.: Hilliard & Brown; Boston: Hilliard, Gray, Little, & Wilkins, 1832), republished as *A Popular Handbook of the Ornithology of Eastern North America,* revised and annotated by Montague Chamberlain (Boston: Little, Brown, 1896), p. xxii.
53. William Cooper, "Notices of Big-Bone Lick," *Monthly American Journal of Geology* 1 (1831–32):29.
54. James E. DeKay, *Anniversary Address on the Progress of the Natural Sciences in the United States: Delivered before the Lyceum of Natural History of New-York* (New York: C. and G. Carwell, 1826), reprinted in John C. Burham, ed., *Science in America* (New York: Holt, Rinehart and Winston, 1971), p. 81.
55. DeWitt Clinton, "An Introductory Discourse Delivered on the Tenth of May, 1814," *Transactions of the Literary and Philosophical Society of New York* 1 (1815):57–58, 114–16.
56. Edward Quinn, "Thomas Jefferson and the Fossil Record," *Bios* 47 (December 1976):161–64. Jefferson considered his *Megalonyx* and Cuvier's *Megatherium* to be identical. Each is now classified as a type genus for the two families *Megalonychidae* and *Megatheriidae.*
57. Thomas Jefferson, "A Memoir on the Discovery of Certain Bones of a Quadruped of the Clawed Kind in the Western Parts of Virginia," *Proceedings of the American Philosophical Society* 4 (1799):246–60.
58. Blane, *Travels,* p. 22.
59. Clinton, "Discourse," p. 27.
60. S. L. Mitchill, *Essay on the Theory of the Earth by M. Cuvier with Mineralogical Notes and an Account of Cuvier's Geological Discoveries by Professor Jameson to Which is Now Added Observations on the Geology of North America* (New York: Kirk and Mercein, 1818). For American response, see [Edward Hitchcock], "The New Theory of the Earth," *North American Review* 28 (1829):265–66; see also the unsigned "Notice and Review of the 'Reliquiae Deluvianae'," *American Journal of Science* 8 (August 1824):151. Familiarity of at least some American readers with Cuvier's zoology is evident in a painstaking but anonymous review of M'Mutrie's American translation of the *Animal Kingdom* in *Monthly American Journal of Geology* 1 (April 1832):447–56; ibid. 2 (June 1832):549–58.
61. Lester P. Coonan and Charlotte M. Porter, "Thomas Jefferson and American Biology," *BioScience* 26 (December 1976):747. See Jefferson, "Memoir," pp. 255–56, and George Turner, "Memoir on the Extraneous Fossils Denominated Mammoth Bones . . . ," *Proceedings of the American Philosophical Society* 4 (1799):510–18, and Caspar Wistar, "A Description of the Bones Deposited by the President," ibid., pp. 526–31.
62. James, *Account* (1823), 2:107.
63. Although *Ursus horribilis* Ord was not captured in the wild, its signs were observed; see James, *Account,* ed. Thwaites, 2:52n–57n; 3:26, 45–51; 4:145–47.

64. James, *Account* (1823), 2:56.
65. Courtney Robert Hall, *A Scientist in the Early Republic* (New York: Columbia University Press, 1934), p. 11.
66. George Catlin, *Letters and Notes on the Manners, Customs and Conditions of the North American Indians,* 4th ed., 2 vols. (London: Published for the author by David Bogue, 1844), 2:361–62.
67. Compare Nuttall's attitudes with those of the father of his friend William Bartram; see Edmund Berkeley and Dorothy Smith Berkeley, *The Life and Travels of John Bartram* (Tallahassee: University Presses of Florida, 1982), pp. 308–18.

Chapter 3: Drawn from Nature

1. See Wilson's illustration of himself for the *Port Folio,* 3d ser., 2 (1809), facing p. 147.
2. James Wilson, "Wilson's *American Ornithology,*" *Blackwood's Edinburgh Magazine* 30 (August 1831):264.
3. See Nathan B. Fagin, *William Bartram: Interpreter of the American Landscape* (Baltimore: Johns Hopkins University Press, 1933), pp. 128–62.
4. Alexander Wilson to William Bartram, October 30, 1803, and August 10, 1804, Wilson Papers, Houghton Library, Harvard University, Cambridge, Massachusetts; see also Alexander Wilson, *American Ornithology, or the Natural History of the Birds of the United States,* 9 vols. (Philadelphia: Bradford and Inskeep, 1808–14), 2:viii.
5. For more on Wilson's techniques and sources, see Jared Sparks, "Wilson's and Bonaparte's Ornithology," *North American Review* 24 (January 1827):116; Robert H. Welker, *Birds and Men: American Birds in Science, Art, Literature, and Conservation, 1800–1900* (New York: Atheneum, 1966), pp. 18–47.
6. Wilson, *American Ornithology,* 3:27–28.
7. Alexander Wilson to Alexander Lawson, March 12, 1804, Wilson Papers.
8. Wilson, *American Ornithology,* 3:26.
9. See the ledger pages, Wilson Papers.
10. George Ord, "Biographical Sketch of Alexander Wilson," in Wilson, *American Ornithology,* 9:xliv.
11. Alexander Wilson to Daniel H. Miller, October 12, 1808, Wilson Papers.
12. Ord, "Biographical Sketch," p. xliv.
13. Wilson, *American Ornithology,* 1:59.
14. [Samuel L. Mitchill], "*American Ornithology, or the Natural History of the Birds of the United States,*" *Medical Repository* 14 (1811):48.
15. "*American Ornithology, of the Natural History of the United States,*" *Port Folio* 3 (June 1814):578–79.
16. "*American Ornithology,*" *American Medical and Philosophical Register* 4 (1814):574; Sparks, "Wilson's and Bonaparte's Ornithology," p. 116.
17. Samuel L. Mitchill, "*American Ornithology,*" *Medical Repository* 17 (1814):250.
18. See "*American Ornithology,*" *American Medical and Philosophical Register* 4 (1814):579. For extended recent discussions, see John C. Greene, "American Science Comes of Age, 1780–1820," *Journal of American History* 60 (1968):22–41; Greene, "The Development of Mineralogy in Philadelphia," *Proceedings of*

the *American Philosophical Society* 113 (1969):283–95; George C. Daniels, "The Process of Professionalization in American Science, 1820–1860," *Isis* 58 (1967):151–66; Leonard Wilson, "The Emergence of Geology as Science in the United States," *Journal of World History* 10 (1967):416–37; C. Stuart Gager, "Botanic Gardens in Science and Education," *Science* 85 (1937):393–99.

19. William Maclure, *Observations on the Geology of the United States of America: With Some Remarks on the Effect Produced on the Nature and Fertility of Soils by the Decomposition of Different Classes of Rocks* (1808; enlarged ed., Philadelphia: Printed for the author by Abraham Small, 1817).

20. For comprehensive accounts of these events, see E. S. Dana, ed., *A Century of Science in America with Special Reference to the American Journal of Science* (New Haven: Yale University Press, 1918), pp. 20–36; Herman Neale Fairchild, *History of the New York Academy of Sciences* (New York: Published by the author, 1887), pp. 1–20; and the unsigned "The New York Academy of Sciences and the American Intellectual Tradition: An Historical Overview," *Transactions of the New York Academy of Sciences,* 2d ser., 37 (1975):3–13.

21. For discussions of Wilson's impact, see *Notice of the Academy of Natural Sciences of Philadelphia* (Philadelphia: Mifflin and Parry, 1831), p. 4; Patsy Gerstner, "The Academy of Natural Sciences of Philadelphia, 1812–1850," in Alexandra Oleson and Sanborn C. Brown, eds., *The Pursuit of Knowledge in the Early American Republic* (Baltimore: Johns Hopkins University Press, 1976), pp. 174, 186.

22. Thomas Jefferson to the Rev. Joseph Willard, March 24, 1789, quoted by Charles A. Browne, "Thomas Jefferson and the Scientific Trends of His Time," *Chronica Botanica* 8 (1944):383.

23. For early expressions of their sentiments, see Harry B. Weiss and Grace M. Ziegler, *Thomas Say, Early American Naturalist* (Baltimore: Charles C. Thomas, 1931), pp. 36–39; C. S. Rafinesque to George Ord, October 1, 1817, T. J. Fitzpatrick Collection, Spencer Library, University of Kansas, Lawrence, Kansas.

24. Quoted by James Southall Wilson, *Alexander Wilson, Poet: Naturalist* (New York: Neale, 1906), pp. 79–80; also p. 85; see pp. 78–89 for their entire correspondence.

25. William Bartram to Thomas Jefferson, March 18, 1805, in Helen G. Cruickshank, ed., *John and William Bartram's America* (New York: Devin-Adair, 1957), pp. 369–70.

26. Wilson, *American Ornithology,* 1:66, 50–51, 155.

27. Ibid., pp. 153, 104–5, 157.

28. David Ramsay, "An Oration on the Advantages of American Independence," *United States Magazine* 1 (1799):53.

29. George Ord, "North American Zoology," in William Guthrie, ed., *A New Geographical, Historical and Commercial Grammar* (Philadelphia: Johnson & Walker, 1815; rpt. Haddonfield: Samuel N. Rhoads, 1894), p. 313.

30. Charles Lucien Bonaparte, *American Ornithology; or, The Natural History of Birds Inhabiting the United States Not Given by Wilson,* 4 vols. (Philadelphia: S. A. Mitchill; Carey, Lea & Carey, 1825–34), 3:7–8.

31. Thomas Say to Thaddeus W. Harris, July 30, 1825, Harris Papers, Houghton Library, Harvard University.

32. Thomas Say to Thaddeus W. Harris, November 28, 1830, ibid.

Chapter 4: Writing the Book of Nature

1. Richard Harlan, *Fauna Americana, Being a Description of the Mammiferous Animals Inhabiting North America* (Philadelphia: Anthony Finley, 1825), p. ix.

2. The impact of Frederick Pursh's "pirating" on American botanists is discussed by Jeannette E. Graustein, *Thomas Nuttall, Naturalist* (Cambridge, Mass.: Harvard University Press, 1967), pp. 85–91.

3. Thomas Say to Thaddeus W. Harris, May 20, 1830, Harris Papers, Houghton Library, Harvard University.

4. For example, see [Samuel Latham Mitchill], *"American Ornithology, or the Natural History of the Birds of the United States,"* *Medical Repository* 14 (1811):48.

5. Gilbert Chinard, "The American Sketchbooks of Charles-Alexandre LeSueur," *Proceedings of the American Philosophical Society* 93 (1949):116–118; Graustein, *Thomas Nuttall*, pp. 7–12.

6. Harry B. Weiss and Grace M. Ziegler, *Thomas Say, Early American Naturalist* (Baltimore: Charles C. Thomas, 1931), p. 39; see also Thomas Say to John F. Melsheimer, April 27, 1817, ibid., pp. 50–51.

7. Jared Sparks, "Wilson's and Bonaparte's Ornithology," *North American Review* 24 (1827):114; James Wilson, "Audubon's Ornithological Biography," *Blackwood's Magazine* 30 (July 1831):11; Charles Lucien Bonaparte, *American Ornithology; or the Natural History of Birds Inhabiting the United States Not Given by Wilson,* 4 vols. (Philadelphia: S. A. Mitchell; Carey, Lea & Carey, 1825–34), 1:iii.

8. Bonaparte, *American Ornithology,* 1:v–vi; see also [Richard Harlan], *Refutation of Certain Misrepresentations, Issued against the Author of the "Fauna Americana" in the Philadelphia Franklin Journal No. 1, 1826 and in the North American Review, No. 50* (Philadelphia: William Stavely, 1826), p. 10.

9. William Keating, *Narrative of an Expedition to the Source of St. Peter's River, Lake Winnepeck, Lake of the Woods, &c. &c.* (Philadelphia: H. C. Carey and I. Lea, 1824), p. 379. See J. Percy Moore, "William Maclure—Scientist and Humanitarian," *Proceedings of the American Philosophical Society* 91 (1947):234–49.

10. William Maclure, "Observations on the Geology of the United States, Explanatory of a Geologic Map," *Transactions of the American Philosophical Society* 6 (1809):411–28, enlarged and revised as *Observations on the Geology of the United States of America* (Philadelphia: Abraham Small, 1817), pp. iii–iv.

11. See "Extracts from Letters Addressed to the Editor [Benjamin Silliman] by William Maclure," *American Journal of Science* 9 (1825):150–61, dated September 10, 1824.

12. Arthur E. Bestor, ed., "Education and Reform at New Harmony: Correspondence of William Maclure and Marie DuClos Frétageot, 1820–1833," *Indiana Historical Society Publication* 15 (1948):371, 388, 401–2, 405.

13. [Samuel G. Morton], *Notice of the Academy of Natural Sciences of Philadelphia,* 2d ed. (Philadelphia: Mifflin and Parry, 1831), p. 5.

14. Thomas Say, *American Entomology, or Descriptions of the Insects of North America,* 3 vols. (Philadelphia: Samuel Augustus Mitchell, 1824, 1825, 1828), unpaginated.

15. Jessie Poesch, *Titian Ramsay Peale and His Journals of the Wilkes Expedition* (Philadelphia: American Philosophical Society, 1961), p. 60.

16. John Godman, *American Natural History*, 3 vols. (Philadelphia: H. C. Carey and I. Lea, 1826–28).

17. Samuel G. Morton, *Crania Americana, or a Comparative View of the Skulls of Various Aboriginal Nations of North and South America* (Philadelphia: J. Dobson, 1839); Thomas Nuttall, *The North American Sylva of F. Andrew Michaux,* vol. 1 (Philadelphia: J. Dobson, 1842); vols. 2 and 3 (Philadelphia: Robert Smith, 1852 [actually printed in 1849]).

18. Say, *American Entomology*, vol. 3, unpaginated text to plate 40.

19. The failure of expeditions to meet public expectations was to be a source of great bitterness for Titian Peale; see his "The South Seas Surveying and Exploring Expedition, the Organization, Equipment, Purposes, Results, and Termination," typed ms. copy, ca. 1855, pp. 10, 16, 26, American Museum of Natural History, New York.

20. Edwin James, *Account of an Expedition from Pittsburgh to the Rocky Mountains, Performed in the Years 1819, 1820,* 2 vols. (Philadelphia: H. C. Carey and I. Lea, 1823), 1:78.

21. Say, *American Entomology*, vol. 2, unpaginated text to plate 20.

22. See Thomas Say, *Glossary to Say's Entomology* (Philadelphia: S. A. Mitchell, 1825); Say, *Glossary to Say's Conchology* (New Harmony, Ind.: Richard Beck and James Bennett, 1832).

23. Thomas Nuttall, *The Genera of North American Plants and Catalogue of the Species to the Year 1817,* 2 vols. (Philadelphia: D. Heartt, 1818); Nuttall, *North American Sylva,* 1:59.

24. Nuttall, *North American Sylva,* 1:47.

25. George Ord, "A Memoir of Thomas Say," in John L. LeConte, ed., *The Complete Writings of Thomas Say on the Entomology of North America,* 2 vols. (New York: Baillière Brothers, 1859), 1:xvi–xvii.

26. See the second revised edition of Thomas Nuttall, *A Popular Handbook of Ornithology of Eastern North America by Thomas Nuttall,* ed. Montague Chamberlain, 2 vols. (Boston: Little, Brown, 1896), 1:vii–viii.

27. Bonaparte, *American Ornithology,* 1:6, 38, 49.

28. See Alexander Garden to John Ellis, May 6, 1757, in James E. Smith, ed., *A Selection of the Correspondence of Linnaeus and Other Naturalists,* 2 vols. (London: Longman, Hurst, Rees, Orme, and Brown, 1821), 1:396–97.

29. Stephen H. Long, "Observations on the Geology of the Country Traversed by the Expedition," in James, *Account,* 2:390.

30. William Coleman, *Biology in the Nineteenth Century: Problems of Form, Function, and Transformation* (New York: John Wiley & Sons, 1971), pp. 22–24.

31. Edwin James, *Account of an Expedition from Pittsburgh to the Rocky Mountains Performed in the Years 1819, 1820,* vols. 15–17 in Reuben Gold Thwaites, ed., *Early Western Travels* (Cleveland: Arthur H. Clark, 1905) 15:233.

32. See John C. Dalton's review in the *American Journal of Medical Science,* n.s., 32 (1856): 397, quoted by James H. Cassedy, "The Microscope in American Medical Science, 1840–1860," *Isis* 67 (1976):76; see also p. 78.

33. John Godman, "Rambles of a Naturalist," added posthumously to the third edition of his *American Natural History,* 2 vols. (Philadelphia: R. W. Pomeroy, 1842) 2:298.

34. See Titian Ramsay Peale's prospectus for his *Lepidoptera Americana: Or Origi-*

nal Figures of Moths and Butterflies of North America in Their Various Stages of Existence and the Plants on Which They Feed (Philadelphia: William P. Gibbons, 1833) unpaginated text to plate 7, Plexippus Butterfly (the monarch).

35. C. S. Rafinesque, *Continuation of a Monograph of the Bivalve Shells of the River Ohio* (Philadelphia, 1831), p. 1.
36. Thomas Say, "Conchology," in William Nicholson, ed., *American Edition of the British Encyclopedia or Dictionary of Arts and Sciences,* vol. 4 (Philadelphia: Mitchell, Ames, and White, 1819), unpaginated.
37. [Morton], *Notice,* p. 11; compare with the description of the cabinet by Samuel G. Morton, "History of the Academy of Natural Sciences of Philadelphia," *American Quarterly Register* 13 (1841):436.
38. Say, *American Entomology,* vol. 1, unpaginated text to plate 5.
39. Ibid., vol. 2, unpaginated text to plate 28; vol. 1, unpaginated text to plate 4.
40. See letter of Thomas Jefferson to Dr. John Manners, February 22, 1818, in Edmund H. Fulling, "Thomas Jefferson: His Interest in Plant Life as Revealed in His Writings, II," *Bulletin of the Torrey Botanical Club* 72 (May 1945):249–50. Wilson acknowledged his debt to Jefferson in the *American Ornithology,* 1:8.
41. Bonaparte, *American Ornithology,* 2:2, 5.
42. Bonaparte, *American Ornithology,* 2:1–5. See, for example, the female gold crest in Bonaparte, *American Ornithology,* vol. 1, plate 2, fig. 4.
43. See the *American Ornithology: Constable's Miscellany of Original and Selected Publications in Various Departments of Literature, Science, and the Arts* (Edinburgh: Constable, 1831), vol. 71.
44. See, for example, "The *American Ornithology* of Alexander Wilson and Charles Lucien Bonaparte," *Edinburgh Evening Post,* November 2, 1832.
45. Charles Winterfield [pseud. for C. W. Weber], "American Ornithology," *American Review* 1 (March 1845):263.
46. Bonaparte, *American Ornithology,* 2:3; 1:52.
47. Nuttall, *Genera,* 1:vi.
48. See Say's proposal in the *New Harmony Disseminator,* October 17 and 24, 1827, quoted by Weiss and Ziegler, *Thomas Say,* pp. 139–40.
49. In this regard, drawings by Wilson, LeSueur, Titian Peale, and Lucy Say might well be contrasted with Vesalius-styled medical plates; see, for example, the plate of *Simis concolor* Harlan, *Journal of the Academy of Natural Sciences* 5, pt. 2 (1825).
50. Bonaparte, *American Ornithology,* 1:3–4.
51. Cited by Francis H. Herrick, *Audubon the Naturalist,* 2 vols. (New York: D. D. Appleton, 1917), 1:330. Compare Peale's illustration with Audubon's first published effort, plate iv of Bonaparte's *American Ornithology,* vol. 1.

Chapter 5: The Denizens of Nature

1. [Edward Hitchcock], "The New Theory of the Earth," *North American Review* 28 (1829):266.
2. From William Bartram's planned address to the U.S. Congress, ms., Historical Society of Pennsylvania, cited by Nathan B. Fagin, *William Bartram:*

Interpreter of the American Landscape (Baltimore: Johns Hopkins University Press, 1933), pp. 16–17.

3. John Kirk Townsend, *Narrative of a Journey across the Rocky Mountains to the Columbia River, and a Visit to the Sandwich Islands, Chili, &c.* (Philadelphia: Henry Perkins, 1839), pp. 332–33, in Reuben Gold Thwaites, ed., *Early Western Travels*, vol. 21 (Cleveland: Arthur H. Clark, 1905), p. 170; his opinion was shared by John B. Wyeth, *Oregon: Or a Short History of a Long Journey* (Cambridge: Printed for John B. Wyeth, 1833; reproduced by University Microfilms, Ann Arbor, 1966), p. 84. John was the brother of the group's leader, Nathaniel, creator of the "Nat-wyethium," better known today as the prairie schooner.

4. Georges Louis Leclerc, comte de Buffon, *Histoire naturelle, générale et particulière*, 15 vols. (Paris: L'imprimerie royale, 1749–67), 9:109–10, 114–25, 103–5; compare to Thomas Jefferson, *Notes on the State of Virginia* (Philadelphia: R. T. Rawle, 1801), p. 42.

5. Thomas Nuttall, *A Popular Handbook of the Ornithology of Eastern North America*, ed. Montague Chamberlain, 2d rev. ed., 2 vols. (Boston: Little, Brown, 1896), 1:xxvi. Contrast Nuttall's view with those of his British contemporary, William Buckland, *"The Birds of America," Quarterly Review* 47 (March–July 1832):340.

6. Thomas Nuttall, *Journal of Travels into the Arkansa Territory during the Year 1819* (Philadelphia: Thomas H. Palmer, 1821), in Reuben Gold Thwaites, ed., *Early Western Travels*, vol. 13 (Cleveland: Arthur H. Clark, 1905), p. 187.

7. For an extended discussion, see Daniel J. Boorstin, *The Lost World of Thomas Jefferson* (New York: Henry Holt, 1948), pp. 60–80.

8. Townsend, *Narrative*, p. 333.

9. Gordon R. Willey and Jeremy A. Sabloff, *A History of American Archaeology* (San Francisco: W. H. Freeman, 1974), pp. 29, 30–34; for examples, see Thaddeus M. Harris, *The Journal of a Tour into the Territory Northwest of the Alleghany Mountains . . . 1803*, in Reuben Gold Thwaites, ed., *Early Western Travels*, vol. 3 (Cleveland: Arthur H. Clark, 1904), p. 362; Nuttall, *Journal of Travels*, p. 115.

10. Jefferson, *Notes*, p. 87; M. F. Ashley Montagu, *Edward Tyson, M.D., F.R.S. (1650–1708) and the Rise of Comparative Anatomy in England* (Philadelphia: American Philosophical Society, 1943), pp. 290, 406, 413–14; Richard Harlan, "Description of an Hermaphrodite Orang Outang, Lately Living in Philadelphia," *Journal of the Academy of Natural Sciences* 5, pt. 2 (1825–27):229–36; Samuel G. Morton, *Brief Remarks on the Diversities of the Human Race and on Some Kindred Subjects* (Philadelphia: Messchew and Thompson, 1842), pp. 12–13; C. W. Peale, "Additions and Donations," unidentified newspaper clipping, Academy of Natural Sciences.

11. A. O. Weese, ed., "The Journal of Titian Ramsay Peale, Pioneer Naturalist," *Missouri Historical Review* 41 (1947):160, entry for June 12, 1819.

12. Edwin James, *Account of an Expedition from Pittsburgh to the Rocky Mountains Performed in the Years 1819, 1820*, in Reuben Gold Thwaites, ed., *Early Western Travels*, vol. 15 (Cleveland: Arthur H. Clark, 1905), 15:117, ibid., 14:305, 19, 68, 7.

13. Roger L. Nichols, ed., *The Missouri Expedition (1818–1820)* (Norman: University of Oklahoma Press, 1969), p. 82.

14. Harlan M. Fuller, ed., *The Journal of Captain R. Bell* (Glendale: Arthur H. Clark, 1957), p. 256, entry for August 31, 1820.

15. Nuttall, *Journal of Travels,* p. 258.

16. J. C. Beltrami, *A Pilgrimage in America* (1828; rpr. Chicago: Quadrangle Books, n.d.), p. 41.

17. Jefferson, *Notes,* pp. 139–40.

18. Charles Coleman Sellers, *Mr. Peale's Museum* (New York: W. W. Norton, 1980), p. 53.

19. For an extended discussion see William Stanton, *The Leopard's Spots: Scientific Attitudes toward Race in America, 1815–1859* (Chicago: University of Chicago Press, 1960), pp. 6, 13; see also Benjamin Rush, "Observations Intended to Favor a Supposition That the Black Color (as It Is Called) of the Negroes Is Derived from the LEPROSY," *Transactions of the American Philosophical Society* 4 (1799):289–97.

20. Samuel Latham Mitchill, *The Picture of New York, or, the Traveller's Guide through the Commercial Metropolis of the United States* (New York: Riley, 1807), pp. 93, 34–35.

21. Courtney Robert Hall, *A Scientist in the Early Republic* (New York: Columbia University Press, 1934), p. 115.

22. Frank Spencer, "Two Unpublished Essays on the Anthropology of North America by Benjamin Smith Barton," *Isis* 68 (December 1977):569.

23. See Samuel G. Morton, *Catalogue of Skulls of Man, and the Inferior Animals in the Collection of Samuel G. Morton* (Philadelphia: Turner and Fisher, 1840), pp. 16–18, 34–37, 47.

24. See Morton's manuscripts on these subjects, American Philosophical Society, Philadelphia, Pennsylvania.

25. Samuel G. Morton, *Crania Americana, or a Comparative View of the Skulls of Various Aboriginal Nations of North and South America* (Philadelphia: J. Dobson, 1839), p. 207.

26. See Charles D. Meigs, *A Memoir of Samuel George Morton* (Philadelphia: T. K. & P. G. Collins, 1851), pp. 23–28. For Owen's interests, see R. J. Cooter, "Phrenology: The Provocation of Progress," *History of Science* 14 (1976):213, 227.

27. See also Stanton, *Leopard's Spots,* pp. 45–53.

28. Their debate can be followed through the pages of John Bachman, "An Investigation of the Cases of Hybridity in Animals on Record," *Charleston Medical Journal and Review,* March 1850, pp. 168–97; Samuel G. Morton, *Letter to John Bachman on the Question of Hybridity in Animals* (Charleston: Walker & Johnson, 1850); and John Bachman, *A Notice of the Types of Mankind . . .* (Charleston: Williams & Gettsinger, 1854).

29. See Samuel G. Morton, *Brief Remarks on the Diversities of the Human Race and Some Kindred Subjects* (Philadelphia: Messchew & Thompson, 1842), pp. 9, 11, 21ff.

30. See Charles Pickering, *Races of Man: And Their Geographical Distribution* (Boston: Charles D. Little & James Brown, 1848), p. 248. Pickering's presentation copy to Morton can be seen at the Academy of Natural Sciences.

31. Edward Lurie, "Louis Agassiz and the Races of Man," in Nathan Reingold, ed., *Science in America since 1820* (New York: Science History Publications, 1976), pp. 148–54. See also A. Hunter Dupree, *Asa Gray* (Cambridge, Mass.: Belknap Press of Harvard University Press, 1959), pp. 220, 228–32, 264ff.; George Daniels, ed., *Darwinism Comes to America* (Waltham: Blaisdell, 1968), pp. 1–28; Gertrude Himmelfarb, *Darwin and the Darwinian Revolution* (New York: W. W. Norton, 1962), pp. 295, 395; Cynthia Eagle Russett, *Darwin in America* (San Francisco: W. H. Freeman, 1976), pp. 9–10.
32. C. S. Rafinesque, *On Botany (1820)*, ed. Charles Boewe (Frankfort: Whippoor-will Press, 1983), p. 10, discussing "the useful cane, or *Miegia arundinaria*."
33. Townsend, *Narrative*, p. 333.
34. Rafinesque, *On Botany*, p. 7. His drawings are among the Rafinesque Papers, Transylvania University, Lexington, Kentucky.
35. See Deborah J. Warner, "Science Education for Women in Antebellum America," *Isis* 69 (March 1978):58.
36. For different interpretations, see A. Hunter Dupree, "Thomas Nuttall's Controversy with Asa Gray," *Rhodora* 54 (December 1952):293–303; Patsy Gerstner, "Vertebrate Paleontology, an Early Nineteenth Century Trans-atlantic Science," *Journal of the History of Biology* 3 (1970):137–48.

Chapter 6: The Natural Plan

1. George Sumner, *A Compendium of Physiological and Systematic Botany* (Hartford: Oliver D. Cooke, 1820), pp. 19–20.
2. Amos Eaton to John Torrey, August 4, 1818, Torrey Papers, New York Botanical Gardens, Bronx, New York.
3. John Torrey to William Darlington, May 23, 1835, Darlington Papers, New-York Historical Society, New York.
4. Ethel M. McAllister, *Amos Eaton, Scientist and Educator, 1776–1842* (Philadelphia: University of Pennsylvania Press, 1941), pp. 147, 180–204; see also Markes E. Johnson, "Geology and Early American Reforms in Education: The Rensselaer and New Harmony Schools" (Ph.D. dissertation, Williams College, 1980); Gerald M. Friedman, " 'Gems' from Rensselaer," *Earth Sciences History* 2 (1983):97–102.
5. Amos Eaton, *A Manual of Botany for North America*, 6th ed. (Albany: Oliver Steele, 1833), p. v, and pt. 2, p. 3. See also letters to John Torrey from C. S. Rafinesque, March 2, 1832, and Thomas Nuttall, September 1838, Torrey Papers. For criticism of Eaton, see George Featherstonhaugh, "Eaton's Geology," *Monthly American Journal of Geology* 1 (August 1831):82–90; for a recent reevaluation, see George Perkins Merrill, *The First One Hundred Years of American Geology* (New Haven: Yale University Press, 1924), pp. 77–79, 128–33.
6. Almira Lincoln, *Familiar Lectures on Botany* (New York: F. J. Huntington and Mason and Law, 1852).
7. For an interesting discussion, see François Delaporte, *Nature's Second Kingdom: Explorations of Vegetality in the Eighteenth Century,* trans. Arthur Goldhammer (Cambridge, Mass.: MIT Press, 1982), pp. 9–29, 88.
8. Kentwood D. Wells, "Sir William Lawrence (1783–1867): A Study of Pre-Darwinian Ideas on Heredity and Variation," *Journal of the History of Biology* 4

(Fall 1971):321. See also Richard Burckhardt, *The Spirit of System: Lamarck and Evolutionary Biology* (Cambridge, Mass.: Harvard University Press, 1977), pp. 22–23, 227.

9. Robert McCracken Peck, "Books from the Bartram Library," in Alwyne Wheeler, ed., *Contributions to the History of North American Natural History* (London: Society for the Bibliography of Natural History, 1983), p. 47.

10. For extended discussion, see Ann Leighton, *American Gardens in the Eighteenth Century* (Boston: Houghton Mifflin, 1976), pp. 75–101.

11. For specific examples, see Julia Lewis Morris, *From Seed to Flower* (Philadelphia: Pennsylvania Horticultural Society, 1976), pp. 1–55.

12. Benjamin S. Barton, *Elements of Botany*, 2 vols. (Philadelphia: Printed for the author, 1803), 1:44.

13. Alexander Garden to John Ellis, May 6, 1757, in James Edward Smith, ed., *A Selection of the Correspondence of Linnaeus and Other Naturalists,* 2 vols. (London: Longman, Hurst, Rees, Orme, and Brown, 1821), 1:369.

14. Thomas Say to John F. Melsheimer, November 6, 1817, in Harry B. Weiss and Grace M. Ziegler, *Thomas Say, Early American Naturalist* (Baltimore: Charles C. Thomas, 1831), p. 51, from the original, Historical Society of Pennsylvania.

15. Thomas Jefferson, *Notes on the State of Virginia* (Philadelphia: R. T. Rawle, 1801), pp. 49–141.

16. For a contemporary's view of Bartram's contributions, see [James Wilson], "Review of Wilson's Ornithology," ms. from the library of Sir William Jardine, Alexander Wilson Papers, Houghton Library, Harvard University. For recent evaluation, consult the commentary and annotated index in Francis Harper, ed., *The Travels of William Bartram: Naturalist's Edition* (New Haven: Yale University Press, 1958), pp. 335–667.

17. Francis H. Herrick, *Audubon, the Naturalist,* 2 vols. (New York: D. Appleton, 1917), 2:214–15.

18. Charles Lucien Bonaparte, *American Ornithology; or the Natural History of Birds Inhabiting the United States Not Given by Wilson,* 4 vols. (Philadelphia: S. A. Mitchell; Carey, Lea, & Carey, 1825–32), 1:iii–iv.

19. Melville H. Hatch, "Coleoptera," in *A Century of Progress in the Natural Sciences, 1853–1953* (San Francisco: California Academy of Sciences, 1955), p. 557.

20. Jeannette Graustein, *Thomas Nuttall, Naturalist* (Cambridge, Mass.: Harvard University Press, 1967), p. 161.

21. See Arthur J. Cain, "Logic and Memory in Linnaeus' System of Taxonomy," *Proceedings of the Linnaean Society of London* 169 (1956–57):144–63.

22. Thomas Say, *Glossary to Say's Entomology* (Philadelphia: S. A. Mitchell, 1825), contains thirty-seven pages of definitions. His *Glossary to Say's Conchology* (New Harmony, Ind.: Richard Bein and James Bennett, 1832), pp. 24–25, gives rules for both diminutive compounds and pronunciations. See also Charles L. Bonaparte, "Observations on the Nomenclature of Wilson's Ornithology," *Journal of the Academy of Natural Sciences* 3, pt. 2 (1824):349–71; 4, pt. 1 (1824):25–67, 163–200; 4, pt. 2 (1824):251–78; 5, pt. 1 (1825–27):57–106, continued as "Further Additions to the Ornithology of the United States and Observations on the Nomenclature of Certain Species" and "The Genera

of North American Birds and a Synopsis of the Species Found within the Territory of the United States," *Annals of the Lyceum of Natural History* 2 (1826–28):154–61, 7–128, 293–451; as well as Thomas Nuttall's forgotten textbook *An Introduction to Systematic and Physiological Botany* (Cambridge: Hilliard and Brown, Booksellers to the University, 1827; 2d ed., 1830).

23. Alexis de Tocqueville, *Democracy in America,* trans. George Lawrence (New York: Doubleday, 1969), p. 410.

24. See S. S. Haldeman, *On the Impropriety of Using Vulgar Names in Zoology* (Philadelphia: Carey and Hart, Judiah Dobson and John Pennington; New York: Wiley and Putnam, 1843), pp. 7–13, 34.

25. C. S. Rafinesque, *Flora Telluriana* (Philadelphia: Printed for the author by H. Probasco, 1836), pp. 42, 8 (see also p. 9), 69–70 (see also pp. 14, 35, 41).

26. C. S. Rafinesque, *Herbarium Rafinesquiarum,* Prodromus (Philadelphia: S. C. Beck for the author, 1833), p. 3.

27. Rafinesque, *Flora Telluriana,* p. 15; see also C. S. Rafinesque to John Torrey, May 1, 1819, Torrey Papers, New York Botanical Gardens, Bronx, New York; C. S. Rafinesque, "Fragments of a Letter to Mr. Bory St. Vincent at Paris," *Western Minerva* 1 (1821):72.

28. John Lindley, *The Vegetable Kingdom,* 3d ed. (London: Bradbury & Evans, 1853), p. xxvii, replaced circles with spheres.

29. See also Julius von Sach, *History of Botany,* trans. Henry E. F. Garnsey, rev. J. B. Balfour (Oxford: Clarendon Press, 1906), pp. 129, 137; R. J. Harvey-Gibson, *Outlines of the History of Botany* (London: A. & C. Black, 1919), p. 74.

30. Rafinesque, *Flora Telluriana,* p. 35.

31. C. S. Rafinesque to John Torrey, April 22, 1830, Torrey Papers.

32. His great universal law of Perpetual Mutability is described in "Extract of a letter to Dr. Torrey of New York, dated 1st December 1832," in *Herbarium Rafinesquiarum,* Prodromus, pp. 11–12, and in his "Principles of the Philosophy on New Genera and New Species of Plants and Animals," *Atlantic Journal* 1 (1832–33):163–64. The "Extract" is discussed by E. D. Merrill, "A Generally Overlooked Rafinesque Paper," *Proceedings of the American Philosophical Society* 86 (1942):74.

33. Most of Rafinesque's ideas were ignored. See Dirk Struik, *Yankee Science in the Making* (Boston: Little, Brown, 1948), p. 270, and Edwin M. Betts, "The Correspondence between Constantine Samuel Rafinesque and Thomas Jefferson," *Proceedings of the American Philosophical Society* 87 (1943):368–80, esp. letters of September 9, 1819, November 6, 1820, and January 25, 1821.

34. Alexander Wilson, *American Ornithology, or The Natural History of the Birds of the United States,* 9 vols. (Philadelphia: Bradford and Inskeep, 1808–14), 1:105; Thomas Jefferson, "A Memoir on the Discovery of Certain Bones of a Quadruped of the Clawed Kind in the Western Parts of Virginia," *Proceedings of the American Philosophical Society* 4 (1799):255–56.

35. Compare also Wilson, *American Ornithology,* 1:65; 4:44; 7:86; and Bonaparte, *American Ornithology,* 2:19–20.

36. Thomas Nuttall, *The Genera of North American Plants and Catalague of the Species to the Year 1817,* 2 vols. (Philadelphia: D. Heartt, 1818), 1:128.

37. Thomas Nuttall, *Journal of Travels into the Arkansa Territory during the Year*

1819 (Philadelphia: Thomas H. Palmer, 1821), in Reuben Gold Thwaites, ed., *Early Western Travels,* vol. 13 (Cleveland: Arthur H. Clark, 1905), p. 78.

38. C. S. Rafinesque, *New Flora and Botany of North America* (1836; rpr. Cambridge, Mass.: Murray, 1946), pt. 1, p. 32.

39. Ibid., p. 23.

40. See Charles Pickering, *Chronological History of Plants: Man's Record of His Own Existence Illustrated through Their Names, Uses, and Companionship* (Boston: Little, Brown, 1879).

41. Charles Pickering, "On the Geographical Distribution of Plants," *Transactions of the American Philosophical Society,* n.s., 3 (1830):277.

42. Rafinesque, *Flora Telluriana,* p. 13.

43. James DeKay, *Anniversary Address on the Progress of the Natural Sciences in the United States: Delivered before the Lyceum of Natural History of New-York* (New York: G. & C. Carwell, 1826), p. 76.

44. R.S.T., "On the Causes Which Retard the Advances of Zoological Knowledge," *Monthly American Journal of Geology* 1 (January 1832):302.

45. Thomas Say to Thaddeus W. Harris, July 30, 1825, Harris Papers, Houghton Library, Harvard University.

46. DeKay, *Anniversary Address,* p. 76.

47. John Godman, *American Natural History,* 3 vols. (Philadelphia: H. C. Carey and I. Lea, 1826–28), 1:xi.

48. Thomas Say, *American Entomology, or Descriptions of the Insects of North America,* 3 vols. (Philadelphia: S. A. Mitchell, 1824–28), vol. 2, text to plate 28; Bonaparte, *American Ornithology,* 1:6.

49. John Torrey to Lewis D. Schweinitz, October 15, 1823, and May 3, 1822, in C. L. Shear and N. E. Stevens, eds., "The Correspondence of Schweinitz and Torrey," *Memoirs of the Torrey Botanical Club* 16 (1915–21):194, 159.

50. John Torrey, *A Flora of the Middle and Northern Sections of the United States; or, a Systematic Arrangement and Description of All the Plants Hitherto Discovered in the United States North of Virginia,* 2 vols. (New York: T. and J. Swords, 1823–24), 1:v. See Benjamin Silliman, *"Flora of the Middle and Northern States,"* *American Journal of Science* 8 (1824):178.

51. See "Notice of New Books," *American Journal of Science* 32 (July 1837):182, 183, a review of Charles Lyell's *Principles of Geology* (Philadelphia: James Kay, Jr., and Brother, 1837), the first American edition based on the fifth and last London edition.

52. For example, see F. W. P. Greenwood, "An Address Delivered before the Boston Society of Natural History," *Boston Journal of Natural History* 1 (May 1834):7–14. The rewards of this new focus were immediately recognized. See "Reports on the Herbaceous Plants and Quadrupeds of Massachusetts, the first by Rev. Chester Dewey—the last by Ebenezer Emmons" and "Report on the Invertebrate Animals of Massachusetts, comprising the Mollusca, Crustacea, Annelida, and Radiata," *American Journal of Science* 41 (1841):378–81. For an extended discussion of the Massachusetts survey, see George P. Merrill, "Contributions to a History of American State Geological and Natural History Surveys," *Bulletin of the U.S. Natural History Museum* 109 (1920):149–58.

53. See Andrew D. Rodgers, *John Torrey: A Story of North American Botany* (New York: Hafner, 1965), p. 110.
54. For example, see C. S. Rafinesque to John Torrey, April 5, 1819, Torrey Papers; A. Hunter Dupree, *Asa Gray* (Cambridge, Mass.: Belknap Press of the Harvard University Press, 1959), p. 57.
55. John Lindley, *An Introduction to the Natural System of Botany or a Systematic View of the Organization, Natural Affinities and Geographic Distributions of the Whole Vegetable Kingdom with a Catalogue of North American Genera of Plants* by John Torrey (London: 1830; New York: G. and C. and H. Carvill, 1831).
56. Examples of the enthusiastic reception are [I. Ray], "DeCandolle's Botany," *North American Review* 38 (1834):61–63; "Notice of New Books," *American Journal of Science* 32 (July 1837):211; Asa Gray, "A Natural System of Botany," *American Journal of Science* 32 (July 1837): 292–303; "Comparative View of the Linnaean and Natural Systems of Botany," *Monthly American Journal of Geology* 1 (March 1832):416–22; Lewis Schweinitz to John Torrey, March 29, 1832, in Shear and Stevens, eds., "Correspondence," p. 268.
57. Amos Eaton, *A Manual of Botany for North America*, 6th ed. (Albany: Oliver Steele, 1833), p. v.
58. C. S. Rafinesque to John Torrey, April 1835, Torrey Papers.
59. Thomas Nuttall to Asa Gray, April 17, 23, 1841, Historic Letter File, Gray Herbarium, Harvard University.
60. Rafinesque pictured himself as an American Linnaeus and initially outlined his *Analyse de la nature* on the plan of the *Systema naturae;* see his *Circular Address on Botany and Zoology* (Philadelphia: Printed for the author by S. Merritt, 1816), p. 15.
61. For Gray's subsequent role in the organization of botany, see Jerry Stannard, "Early American Botany and Its Sources," in Thomas R. Buckman, ed., *Bibliography and Natural History* (Lawrence: University of Kansas Libraries, 1966), p. 89.
62. Amos Eaton to John Torrey, November 4, 1836, Torrey Papers.
63. Nuttall, *Journal of Travels*, p. 249.
64. John Torrey to William Darlington, October 6, 1839, Darlington Papers.
65. Dupree, *Asa Gray*, p. 75. See also Leonard Huxley, ed., *Life and Letters of Joseph Dalton Hooker*, 2 vols. (London: J. Murray, 1918), 2:206.
66. Quoted in Rogers, *Torrey*, p. 81.
67. John Torrey, "Some Account of a Collection of Plants," *Annals of the Lyceum of Natural History* 2 (1826–28):161–254, read on December 11, 1826.
68. Erwin Stresemann, *Ornithology from Aristotle to the Present*, trans. Hans J. Epstein and Cathleen Epstein (Cambridge, Mass.: Harvard University Press, 1975), p. 169; W. Otto Emerson, "A Manuscript of Charles Lucien Bonaparte," *Condor* 7 (January–February 1905):44–47; Nicholas Aylward Vigors, "Observations on the Natural Affinities That Connect the Orders and Families of Birds," *Transactions of the Linnaean Society of London* 14 (1825):468, 509. Compare the similarity of Vigors, p. 397, to Bonaparte, *American Ornithology*, 2:5.
69. Stresemann, *Ornithology*, pp. 163, 165; see also Charles L. Bonaparte, "Betrachtungen über den Species," *Journal für Ornithologie* 4 (July 1856):257–59.

70. See William Lawrence, *Lectures on Physiology, Zoology and the Natural History of Man* (Salem: Foote & Brown, 1828), pp. 233–34. This edition was dedicated to Blumenbach.

Chapter 7: A New Organon

1. Allen G. Debus, *Man and Nature in the Renaissance* (Cambridge: Cambridge University Press, 1978), pp. 116–19; for more on American Baconism see Theodore Dwight Bozeman, *Protestants in an Age of Science: The Baconian Ideal and Antebellum American Religious Thought* (Chapel Hill: University of North Carolina Press, 1977), pp. 1–31; Brooke Hindle, *The Pursuit of Science in Revolutionary America, 1735–1789* (1956; rpt. New York: W. W. Norton, 1974), pp. 190, 214–15; and George H. Daniels, *American Science in the Age of Jackson* (New York: Columbia University Press, 1968), p. 106.
2. Edmund Berkeley and Dorothy Smith Berkeley, *The Life and Travels of John Bartram* (Tallahassee: University Presses of Florida, 1982), pp. 32–34, 63.
3. William E. Wilson, *The Angel and the Serpent* (Bloomington: Indiana University Press, 1984), p. 5.
4. Stanley M. Guralnick, *Science and the Ante-Bellum American College* (Philadelphia: American Philosophical Society, 1975), p. 14.
5. Benjamin Smith Barton, *Fragments of the Natural History of Pennsylvania* (Philadelphia: Printed for the author by Way & Graff, 1799), pt. 1, p. viii.
6. For a more extended discussion of his career, see Francis W. Pennell, "Benjamin Smith Barton as a Naturalist," *Proceedings of the American Philosophical Society* 86 (1942–43):108–22.
7. Barton, *Fragments,* p. 24.
8. Guralnick, *Science and the Ante-Bellum American College,* p. 15.
9. Richard M. Gummere, *The American Colonial Mind and the Classical Tradition* (Cambridge, Mass.: Harvard University Press, 1963), p. 56. For an extended discussion see Stanley M. Guralnick, "Sources of Misconception on the Roles of Science in the Nineteenth-Century American College," *Isis* 65 (1975):352–66.
10. Abraham A. Davidson, *The Eccentrics and Other American Visionary Painters* (New York: E. P. Dutton, 1978), pp. 19–21, 28; see also Maclure's revised *Observations on the Geology of the United States* (Philadelphia: Printed for the author by Abraham Small, 1817), pp. iii–iv.
11. Maclure's participation in Owen's socialistic community is discussed by Charles A. Browne, "Some Relations of the New Harmony Movement to the History of Science in America," *Scientific Monthly* 42 (1936):483–97; see also Samuel G. Morton, *A Memoir of William Maclure* (Philadelphia: T. K. & P. G. Collins, 1841), p. 24; for a sympathetic discussion, see J. Percy Moore, "William Maclure—Scientist and Humanitarian," *Proceedings of the American Philosophical Society* 91 (1947):234–49.
12. Hans Aarsleff, "An Outline of Language-Origins Theory since the Renaissance," *Annals of the New York Academy of Science* 280 (1976):9 , 11, 15; see also Rena Reese, ed., "List of Books and Pamphlets in the Library of the Workingman's Institute, New Harmony, Indiana," New Harmony, March 1809, mimeographed, Widener Library, Harvard University.

13. George Sumner, *A Compendium of Physiological and Systematic Botany* (Hartford: Oliver D. Cooke, 1820), pp. 19–20.
14. Almira Lincoln, *Familiar Lectures on Botany* (New York: F. J. Huntington and Mason and Law, 1852), p. 233.
15. "Officers of the Academy of Natural Sciences, for the Year 1818," *Journal of the Academy of Natural Sciences* 1, pt. 2 (1818): 221.
16. Peale's manuscript is owned by the American Museum of Natural History, New York; details of the trip are provided by Charlotte M. Porter, "Following Bartram's 'Track': Titian Ramsay Peale's Florida Journey," *Florida Historical Quarterly* 61 (April 1983):431–34; see also Thomas Say to John V. Melsheimer, July 30, 1816, Melsheimer Correspondence, Academy of Natural Sciences of Philadelphia; published in W. J. Fox, "Letters from Thomas Say to John F. Melsheimer, 1816–1825," *Entomological News and Proceedings of the Entomological Section of the Academy of Natural Sciences* 12 (1901):233.
17. Andrew Ellicott's *Geological Observations in the Mississippi Valley and Florida, 1796–1800* is discussed by George W. White in James X. Corgan, ed., *The Geological Sciences in the Antebellum South* (University, Ala.: University of Alabama Press, 1982), pp. 9–25.
18. Charlton W. Tebeau, *A History of Florida* (Coral Gables: University of Miami Press, 1971), pp. 112–15.
19. Thomas Say to Jacob Gilliams, January 30, 1818, in Harry B. Weiss and Grace M. Ziegler, *Thomas Say, Early American Naturalist* (Baltimore: Charles C. Thomas, 1931), p. 56.
20. Say's findings are given in his unpaginated article "Conchology," in William Nicholson's American edition of the *British Encyclopedia,* vol. 4 (Philadelphia: Mitchell, Ames & White, 1819); Thomas Say to John F. Melsheimer, June 10, 1818, in Fox, "Letters," pp. 235–36.
21. Morton, *Memoir,* pp. 18–19.
22. See W. S. Monroe, *History of the Pestalozzian Movement in the United States* (Syracuse: Bardeen, 1907), and William Maclure, "Mr. Owen and His Plan of Education," *American Journal of Science* 9 (June 1825):254.
23. "Extracts of Letters to the Editor, Dated at Paris, January 10 and March 14, from William Maclure," *American Journal of Science* 9 (1825):254.
24. "Memoranda, Extracted from a Letter to the Editor, Dated Alicante (Spain) March 6, 1824 from William Maclure," *American Journal of Science* 9 (1825):188.
25. Jeannette Graustein, *Thomas Nuttall, Naturalist* (Cambridge, Mass.: Harvard University Press, 1967), pp. 267–93.
26. Wilson, *The Angel and the Serpent,* p. 152.

Chapter 8: The Vortex of Experiment

1. Karl Arndt, *George Rapp's Harmony Society* (Philadelphia: University of Pennsylvania Press, 1965), pp. 65, 84–86, 231.
2. Peter Vorzimmer, "Darwin, Malthus and the Theory of Natural Selection," *Journal of the History of Ideas* 30 (October–December 1969):527–42; Frank N. Egerton, "Humboldt, Darwin and Population," *Journal of the History of Biology* 3 (1970):325; Joel S. Schwartz, "Charles Darwin's Debt to Malthus and

Edward Blythe," *Journal of the History of Biology* 7 (1974):301–18; see also Thomas Malthus, *An Essay on the Principle of Population* (London: J. M. Dents & Sons Ltd., 1803), Book I, pp. 5–25; Book II, pp. 279–94.

3. See Margaret B. C. Canney, *Robert Owen, 1771–1858, Catalogue of an Exhibition of Printed Books, Held in the Library of the University of London, October–December 1958* (London, 1959), p. 10.

4. Paul Wilhelm, *Travels in North America, 1822–1824,* trans. W. Robert Nitske, ed. Savoie Lottinville (Norman: University of Oklahoma Press, 1973), p. 157.

5. George B. Lockwood, *The New Harmony Communities* (Marion, Ind.: Chronicle Co., 1902), p. 74; Richard W. Leopold, *Robert Dale Owen* (Cambridge, Mass.: Harvard University Press, 1940), p. 20.

6. Arndt, *George Rapp's Harmony Society*, pp. 226–27.

7. Lockwood, *New Harmony Communities*, p. 75.

8. F. A. Michaux, *Travels West of the Allegheny Mountains,* in Reuben Gold Thwaites, ed., *Early Western Travels*, vol. 3 (Cleveland: Arthur H. Clark, 1904), p. 303.

9. Canney, *Owen, Catalogue*, p. 11.

10. A. Hunter Dupree, *Asa Gray* (Cambridge, Mass.: Belknap Press of Harvard University Press, 1959), p. 101; William Stanton, *The Great United States Exploring Expedition of 1838–1842* (Berkeley and Los Angeles: University of California Press, 1975), p. 333.

11. See Robert Owen, *A New View of Society; or, Essays on the Principle of the Formation of the Human Character, and the Application of the Principle to Practice* (London: Cadell and Davies, Strand, 1813); and Canney, *Owen, Catalogue*, p. 11.

12. For extended discussions see Arthur E. Bestor, *Backwoods Utopias: The Sectarian and Owenite Phases of Communitarian Socialism in America, 1663–1829* (Philadelphia: University of Pennsylvania Press, 1950), pp. 94–159; and George Lockwood and Charles Prosser, *The New Harmony Experiment* (New York: Appleton, 1905).

13. "Memoranda, Extracted from a Letter to the Editor, Dated Alicante (Spain) March 6, 1824 from William Maclure," *American Journal of Science* 8 (1824):187–90.

14. "Extracts from Letters Addressed to the Editor by William Maclure," *American Journal of Science* 9 (1825):164, dated November 9, 1824.

15. [Benjamin Silliman], "Mr. Owen and His Plan of Education," *American Journal of Science* 9 (1825):383.

16. "Extracts," p. 160, dated September 10, 1824.

17. Lewis Mumford, *Sticks and Stones,* 2d rev. ed. (New York: Dover, 1955), p. 82.

18. Donald C. Peattie, *Green Laurels* (New York: Literary Guild, 1936), p. 251.

19. Bernard Jaffe, *Men of Science* (New York: Simon & Schuster, 1958), pp. 34–37.

20. Ross F. Lockridge, *The Old Fauntleroy Home* (Published for the New Harmony Memorial Commission by Mrs. Edmund Burke Ball, 1939), p. 46; see also p. 56.

21. William E. Wilson, *The Angel and the Serpent* (Bloomington: Indiana University Press, 1984), p. 154.

22. William Maclure, *Opinions on Various Subjects,* 3 vols. (New Harmony: School Press, 1831–38), 3:491–95; see also his letter to S. G. Morton, April 3, 1839, American Philosophical Society, in Nathan Reingold, ed., *Science in Nineteenth-Century America: A Documentary History* (New York: Hill & Wang, 1964), p. 36.
23. Leopold, *Owen,* pp. 25–44.
24. Bernhardt, Duke of Saxe-Weimar Eisenach, *Reise se hoheit des Herzogs Bernhard zu Sachsen-Weimar-Eisenach dürch Nord-Amerika idem Jahren 1825 und 1826* (Weimar: bei Wilhelm Hoffmann, 1828), pp. 130–55, describes New Harmony. The book, translated as *Travels through North America during the Years 1825 and 1826,* 2 vols. (Philadelphia: Carey, Lea and Carey, 1828), 2:110–11, describes an evening of Byron and the social stratification within the community.
25. Wilson, *The Angel and the Serpent,* illustrates Wright dressed in this outfit opposite p. 147.
26. Bernhardt, *Travels,* 2:109; see also R. W. G. Vail, *The American Sketchbooks of Charles Alexandre LeSueur* (Worcester: American Antiquarian Society, 1938), for illustrations of New Harmony.
27. Bernhardt, *Travels,* 2:115.
28. Alexander Philip Maximilian, Prince of Wied-Neuwied, *Travels in the Interior of North America, 1832–1834,* trans. H. Evans Lloyd, 2 vols. (London: Ackermann, 1843), pp. 74–92, describes New Harmony; see also Reuben Gold Thwaites, ed., *Early Western Travels* (Cleveland: Arthur H. Clark, 1906), 22:180.
29. Bernhardt, *Travels,* 2:117; see also p. 112.
30. The last date held by a library is May 14, 1840 (information courtesy of the General Research and Humanities Division of the New York Public Library).
31. Thomas Say to Benjamin Tappan, August 30, 1827, Tappan Papers, Library of Congress, in Reingold, ed., *Science in Nineteenth-Century America,* p. 34.
32. Thomas Say to Benjamin Lockwood, August 30, 1827, in Reingold, ed., *Science in Nineteenth-Century America,* p. 34; Lockwood and Prosser, *New Harmony Experiment,* p. 243.
33. Marie Frétageot to William Maclure, November 8, 1831, Paris, quoted in Harry B. Weiss and Grace M. Ziegler, *Thomas Say, Early American Naturalist* (Baltimore: Charles C. Thomas, 1931), p. 150.
34. The best source for details of daily community life is Arthur E. Bestor, "Education and Reform at New Harmony: Correspondence of William Maclure and Marie DuClos Frétageot, 1820–1833" *Indiana Historical Society Publication* 15 (1948):285–417.
35. Lucy Say to Arthur F. Gray, February 13, 1883, American Philosophical Society, Philadelphia.
36. Frétageot to Maclure, February 10, 1831, in Weiss and Ziegler, *Say,* p. 146.
37. Frétageot to Maclure, July 28, 1831, quoted in ibid., p. 148.
38. Maclure to Frétageot, December 29, 1826, in Bestor, "Education and Reform," p. 373.
39. Thomas Nuttall, *The Genera of North American Plants and Catalogue of the Species to the Year 1817,* 2 vols. (Philadelphia: D. Heartt, 1818), 2:233.
40. See the Titian R. Peale Papers, American Museum of Natural History, New

York; C. S. Rafinesque, *A Life of Travels* (Philadelphia: Printed for the author, 1836), pp. 78–79; C. S. Rafinesque to Amos Eaton, July 30, 1826, T. J. Fitzpatrick Collection, Spencer Library, University of Kansas, Lawrence.

41. See Constantine S. Rafinesque, *Herbarium Rafinesquianum* (Philadelphia: S. C. Beck, 1833), pp. 3–5, the first part of a catalog printed from the same type as a list published earlier in his *Atlantic Journal* 1 (1834):167–68. For this information I am indebted to Charles Boewe, editor of the C. S. Rafinesque Papers.

42. Benjamin H. Coates, *Biographical Sketch of the Late Thomas Say, Esq.* (Philadelphia: Published by the Academy, 1835), p. 23; see also Harry B. Weiss and Grace M. Ziegler, "The Communism of Thomas Say," *Journal of the New York Entomological Society* 35 (1927):231–39.

43. Say to Harris, January 29, 1829, and July 5, 1830, Harris Papers, Houghton Library, Harvard University.

Chapter 9: Maclure's Legacy

1. See James X. Corgan, "Early American Geological Surveys and Gerard Troost's Field Assistants, 1831–1836," in James X. Corgan, ed., *The Geological Sciences in the Antebellum South* (University, Ala.: University of Alabama Press, 1982), pp. 41–46.

2. Nicollet, who is credited with the first use of fossils to correlate western geological strata, explored parts of Louisiana in 1835 and proceeded to map the Mississippi the next year. In 1838, he was assisted in the Dakota Territory by John Charles Frémont, who extended Maclure's blend of science and politics by leading Missouri governor Thomas Benton's secret expedition to the Rockies.

3. Featherstonhaugh, a British geologist, was the editor of the *Monthly American Journal of Geology,* now a rare journal, the only volume of which (1831–32) is available as a reprint (New York: Hafner, 1969) with introduction by George W. White. See also [George P. Merrill], *"The Monthly American Journal of Geology and Natural Sciences,"* *American Geologist* 30 (July–December 1902):62–64. For more on Featherstonhaugh's biography, see Joan M. Eyles, "G. W. Featherstonhaugh (1780–1866), F.R.S., F.G.S., Geologist and Traveller," *Journal of the Society for the Bibliography of Natural History* 8 (1978):381–95.

4. A. Hunter Dupree, *Science in the Federal Government* (Cambridge, Mass.: Harvard University Press, 1957), pp. 77–78.

5. For a refreshingly amusing account of these events, see George Ord to Charles Waterton, November 21, 1842, typescript from the original MS, private collection, microfilm, American Philosophical Society, Philadelphia.

6. Lewis Mumford, *Sticks and Stones,* 2d rev. ed. (New York: Dover, 1955), pp. 38, 82.

7. Richard W. Leopold, *Robert Dale Owen* (Cambridge, Mass.: Harvard University Press, 1940), pp. 219–26.

8. Ronald Nye, *The Cultural Life of a New Nation* (New York: Harper & Row, 1960), pp. 166–67.

9. Harry B. Weiss and Grace M. Ziegler, *Thomas Say, Early American Naturalist* (Baltimore: Charles C. Thomas, 1931), p. 164.

10. Leopold, *Owen*, pp. 268–83; see also Elinor Pancoast, *The Incorrigible Idealist: Robert Dale Owen in America* (Bloomington: Principia Press, 1940).
11. George B. Lockwood, *The New Harmony Communities* (Marion, Ind.: Chronicle Co., 1902), p. 178.
12. Ross R. Lockridge, *The Old Fauntleroy Home* (Published for the New Harmony Memorial Commission by Mrs. Edmund Burke Ball, 1939), p. 120.
13. Leopold, *Owen*, pp. 47–52.
14. The *Daily Sentinel* began publication on January 10, 1830, and Owen's *Working Man's Advocate* on October 31, 1829. The two were later combined, although the *Advocate* surfaced again in the 1840s (information courtesy of the New York Public Library); see also Pancoast, *Incorrigible Idealist*, p. 115.
15. Dupree, *Science in the Federal Government*, pp. 63, 70–72; see also Dirk Struik, *Yankee Science in the Making* (Boston: Little, Brown, 1948), p. 200; Donald C. Peattie, *Green Laurels* (New York: Literary Guild, 1936), p. 261.
16. Alexander Maclure to S. G. Morton, letter 110, microfilm, American Philosophical Society.
17. Alexander Maclure to George Engelmann, January 13, 1846, Engelmann Papers, Missouri Botanic Garden, St. Louis.
18. Each paper is individually paginated. Joseph Leidy, "A Memoir on the Extinct Sloth Tribe," *Smithsonian Contributions to Knowledge* 7 (1855):1–68; John Chappelsmith, "Account of a Tornado Near New Harmony, Indiana, April 30, 1852," ibid., pp. 3–11; for more on Leidy's reputation see Huxley, ed., *Life and Letters*, 2:208.
19. For more see Lockwood, *New Harmony Communities,* pp. 294–306.
20. This neglected aspect of New Harmony can be traced through the pamphlet and clippings collection in the Owen Family Papers, Manuscript Division, New York Public Library.
21. Jacob Schenck and Richard Owen, *The Rappites: Interesting Notes about New Harmony* (Evansville: Courier Co., 1890), p. 9; see also p. 12 and Leopold, *Owen*, pp. 321–40.
22. Charles A. Browne, "Some Relations of the New Harmony Movement to the History of Science," *Scientific Monthly* 42 (1936):492.
23. In 1812 Eaton, a lawyer, was sentenced to life imprisonment with hard labor on charges involving embezzlement and fraud. He was released from prison in 1815 at the end of the war. During that period, both he and Rafinesque had sons named Charles Linnaeus, who died. See Ethel M. McAllister, *Amos Eaton, Scientist and Educator, 1776–1842* (Philadelphia: University of Pennsylvania Press, 1941), p. 134.
24. William Maclure, *Opinions on Various Subjects*, 3 vols. (New Harmony: School Press, 1831–38), 1:38–39, 40.
25. For an extended discussion, see Ian MacPhail, "Natural History in Utopia: The Works of Thomas Say and François André Michaux Printed at New Harmony, Indiana," in Alwyne Wheeler, ed., *Contributions to the History of North American Natural History* (London: Society for the Bibliography of Natural History, 1983), pp. 16–18. See also "Remarks of the Publishing Committee," *Journal of the Academy of Natural Sciences* 1, pt. 2 (1818):485–86.
26. See *Contributions of the Maclurian Lyceum to the Arts and Sciences* 1 (January 1827):1; copies of this rare serial may be seen in Houghton Library, Harvard University.

27. Asa Gray, "Notice of the Botanical Writings of the Late C. S. Rafinesque," *American Journal of Science* 40 (January–March 1841):221–41, discussed at length by Charlotte M. Porter, " 'Subsilentio': Discouraged Works of American Natural History of the Early Nineteenth Century," *Journal of the Society for the Bibliography of Natural History* 9 (1979):109–19.

28. Draft letter to Lucy Say, March 19, 1835, Harris Papers, Houghton Library.

29. Alexander Philip Maximilian, Prince of Wied-Neuwied, *Travels in the Interior of North America, 1832–1834,* trans. H. Evans Lloyd, 2 vols. (London: Ackermann, 1843), in Reuben Gold Thwaites, ed., *Early Western Travels,* (Cleveland: Arthur H. Clark, 1906), 22:170.

30. George Ord, "A Memoir of Charles-Alexandre LeSueur," *American Journal of Science,* 3d ser., 8 (1849):189–216, provides a useful summary of LeSueur's career.

31. Maximilian, *Travels,* in Thwaites, ed., *Early Western Travels,* 22:170. LeSueur's remarkable notebooks of drawings have been described more recently as the closest approximation which may be found to a snapshot"; see Gilbert Chinard, "The American Sketchbooks of Charles-Alexandre LeSueur," *Proceedings of the American Philosophical Society* 93 (1949):116.

32. Karl Bernhardt, Duke of Saxe-Weimer Eisenach, *Travels through North America during the Years 1825 and 1826,* 2 vols. (Philadelphia: Carey, Lea and Carey, 1828), 2:113.

33. Maclure decided to transfer part of the New Harmony library to the academy in 1835, and the librarian, Charles Pickering, went to New Harmony to accomplish this move; see Samuel G. Morton, *A Memoir of William Maclure* (Philadelphia: T. K., & P. G. Collins, 1841), p. 24.

34. The earliest edition of Nuttall's *Sylva* bears the date 1842, but it was not printed until 1846. Before he departed for England, Nuttall left the manuscript with his publisher, Judah Dobson, whose death and defective letterpress delayed publication; see Elias Durand, "Memoir of the Late Thomas Nuttall," *Proceedings of the American Philosophical Society* n.s., 7 (1861):313.

35. The exquisite plates Maclure purchased for New Harmony were executed by Jean Baptiste Audebert (1759–1800) for Louis Jean Pierre Vieillot's unfinished *Histoire naturelle des oiseaux de l'Amerique septentrionale* (Paris, 1807). In 1865, John Cassin wrote to P. L. Slater that he was authorized by the academy to sell them for scrap.

36. Maclure, *Opinions,* 1:52.

37. Benjamin H. Coates, *Biographical Sketch of the Late Thomas Say, Esq.* (Philadelphia: Published by the Academy, 1835), pp. 13ff.

38. *New Harmony Disseminator,* October 24, 1827, quoted by Weiss and Ziegler, *Say,* pp. 139–40.

39. Thomas Say, *Glossary to Say's Conchology* (New Harmony, Ind.: Richard Beck and James Bennett, 1832).

40. See Say's unsigned article "Conchology" in William Nicholson, ed., *American Edition of the British Encyclopedia or Dictionary of Arts and Sciences,* 12 vols. (Philadelphia: Mitchell, Ames, and White, 1819), vol. 4, unpaginated.

41. C. A. Poulson, "Rafinesque's Monograph of the Bivalve Shells of the River Ohio," *Monthly American Journal of Geology and Natural Sciences* 1 (February 1832):370–77, published in the original French September 1820 in *Annales Générales des Sciences Physiques.*

42. On the first page of his *Continuation of a Monograph of the Bivalve Shells of the River Ohio* (Philadelphia, 1831), Rafinesque states in his by now fluent English that "Mr. Say" knows "so little of the animals of these shells, as to have mistaken their mouths for their tail, and their anterior for the posterior parts of the shells."

43. Thomas Say to Charles Wilkins Short, November 20, 1831, Say MSS, Museum of Comparative Zoology, Harvard University. See also the presentation copy to Thomas Say; C. A. Poulson, *A Monograph of the Fluviatile Bivalve Shells of the River Ohio* (Philadelphia: J. Dobson, 1832).

44. See Thaddeus Harris to Dr. D. H. Storer of Boston, November 2, 1836, cited by Arnold Mallis, *American Entomologists* (New Brunswick: Rutgers University Press, 1971), p. 30.

45. Lucy Say to Thaddeus Harris, March 15, 1835, Harris Papers, Houghton Library.

46. See Lucy Say to Thomas Bland, December 15, 1859, Say MSS, referring to the awaited W. G. Binney, ed., *The Complete Writings of Thomas Say on the Conchology of the United States,* 2 vols. (New York: H. Baillière, 1858), a work of 252 pages with plates. For an evaluation of Say's taxonomy, see V. Sterki, "A Few Notes on Say's Early Writings and Species," *Nautilus* 21 (1907): 31–34.

47. Mallis, *American Entomologists,* pp. 24, 245–48.

48. John L. LeConte, ed., *The Complete Writings of Thomas Say on the Entomology of North America,* 2 vols. (New York: Baillière Bros., 1859), followed by editions in London, Paris, and Madrid.

49. Francis W. Pennell, "Travels and Scientific Collections of Thomas Nuttall," *Bartonia,* no. 18 (December 1936), p. 44.

50. A. Hunter Dupree, *Asa Gray* (Cambridge, Mass.: Belknap Press of Harvard University Press, 1959), pp. 151–54, 211–12.

Chapter 10: The Rise of Peer Review

1. A. O. Weese, ed., "The Journal of Titian Ramsay Peale, Pioneer Naturalist," *Missouri Historical Review* 41 (1947):150, entry for Tuesday, May 4, 1819.

2. Quoted by Theodore Dwight Bozeman, *Protestants in an Age of Science: The Baconian Ideal and Antebellum American Religious Thought* (Chapel Hill: University of North Carolina Press, 1977), p. 47.

3. Nathan Reingold, "Definitions and Speculations: The Professionalization of Science in America in the Nineteenth Century," in Alexandra Oleson and Sanborn C. Brown, eds., *The Pursuit of Knowledge in the Early American Republic* (Baltimore: Johns Hopkins University Press, 1976), pp. 33–69, esp. pp. 48–49, 53.

4. George Ord, "A Memoir of Thomas Say," in John L. LeConte, ed., *The Complete Writings of Thomas Say on the Entomology of North America,* 2 vols. (New York: Baillière Bros., 1859), 2:viii–ix.

5. Thaddeus W. Harris to Samuel G. Morton, April 5, 1838, microfilm of originals owned by Hugh Montgomery, American Philosophical Society, Philadelphia; see also Asa Gray to W. H. Hooker, January 15, 1841, in Jane Loring, ed., *Letters of Asa Gray,* 2 vols. (Boston: Houghton, Mifflin, 1894), 2:278.

6. See Thomas Nuttall to Asa Gray, April 7, 1841, Historic Letter File, Gray Herbarium, Harvard University, Cambridge, Massachusetts. For a different analysis of this exchange, see A. Hunter Dupree, "Thomas Nuttall's Controversy with Asa Gray," *Rhodora* 54 (1952):293–303.

7. Patsy Gerstner, "Vertebrate Paleontology, an Early Nineteenth-Century Transatlantic Science," *Journal of the History of Biology* 3 (1970):137–48.

8. Richard Harlan, *Refutation of Certain Misrepresentations Issued against the Author of "Fauna Americana"* . . . (Philadelphia: William Stavely, 1826), p. 6.

9. H.E.U. (John Godman) to Isaac Hays, August 16 and January 26, 1827, Isaac Hays Correspondence, American Philosophical Society.

10. Thomas Say to Thaddeus W. Harris, July 30, 1825, Harris Papers, Houghton Library, Harvard University.

11. See Anna Maclure to S. G. Morton, April 19, 1842, microfilm, American Philosophical Society; and Samuel G. Morton, *A Memoir of William Maclure* (Philadelphia: T. K. and P. G. Collins, 1841), p. 24.

12. Thomas Say to C. L. Bonaparte, December 12, 1828, microfilm, American Philosophical Society; see also Thomas Say to Thaddeus W. Harris, January 4, 1829, Harris Papers. The titles of these rare papers are *Descriptions of New Species of Curculionites of North America, with Observations on Some of the Species Already Known,* 1831; *Descriptions of New Species of Heteropterous Hemiptera of North America,* December 1831; *Descriptions of New Species of North American Insects and Observations on Some of the Species Already Described,* 1829–33. Copies of these rare printed materials are owned by the American Museum of Natural History, New York City. The library of the U.S. Department of Agriculture owns a copy of the fourth paper, Say's *Descriptions of New Species of North American Insects Found in Louisiana by Joseph Barabino,* March 1831.

13. Thomas Say to C. L. Bonaparte, July 11, 1826, microfilm, American Philosophical Society.

14. Thomas Say to Thaddeus W. Harris, January 4, 1829, Harris Papers.

15. Thomas Say to Thaddeus W. Harris, January 4, 1829, May 20, 1830, ibid.

16. Thomas Nuttall to Asa Gray, April 7, 1841, Historic Letter File, Gray Herbarium.

17. Audubon's errors have been noted by Susan Klimley in the winter 1978 issue of the *North American Newsletter of the Society for the Bibliography of Natural History.* See also Jeannette Graustein, "Audubon and Nuttall," *Scientific Monthly* 74 (1952):88, and her definitive biography, *Thomas Nuttall, Naturalist* (Cambridge, Mass.: Harvard University Press, 1967), pp. 318, 323.

18. John K. Townsend, *Narrative of a Journey across the Rocky Mountains to the Columbia River, and a Visit to the Sandwich Islands, Chili, &c.* (Philadelphia: Henry Perkins, 1839). Samuel G. Morton, *Catalogue of Skulls of Man and the Inferior Animals in the Collection of Samuel G. Morton* (Philadelphia: Turner and Fisher, 1840), pp. 16–18 and 34–36 lists at least thirty-five specimens, some of which were published in Morton's *Crania Americana, or a Comparative View of the Skulls of Various Aboriginal Nations* (Philadelphia: J. Dobson, 1839), after p. 210.

19. See Benjamin Silliman to Heinrich Georg Brown, July 25, 1845, American Philosophical Society.

20. John K. Townsend, *Ornithology of the United States or; Descriptions of the Birds*

Inhabiting the States and Territories of the Union, vol. 1 (Philadelphia: J. B. Chevalier, 1839).

21. Thaddeus W. Harris to Samuel G. Morton, April 5, 1838, microfilm, American Philosophical Society.

22. His generosity is noteworthy because, as Graustein points out (*Nuttall,* pp. 325–26), Washington Irving's novel *Astoria* published in October 1836 gave Nuttall's participation in Hunt's romanticized 1811 expedition wide publicity which the naturalist may have resented.

23. Thomas Nuttall to Asa Gray, April 7, 1841, Historic Letter File, Gray Herbarium.

24. See DeWitt Clinton, *An Introductory Discourse Delivered before the Literary and Philosophical Society of New-York on the Fourth of May, 1814* (New York: David Longworth, 1815), pp. 27–33, 73–76.

25. See Charlotte M. Porter, " 'Subsilentio': Discouraged Works of American Natural History," *Journal of the Society for the Bibliography of Natural History* 9 (1979):109–19.

26. Couthouy's eventual court-martial is discussed by William Stanton, *The Great United States Exploring Expedition of 1838–1842* (Berkeley and Los Angeles: University of California Press, 1975), p. 284.

27. William L. Hudson, Journal, 1:328–29, entry for November 26, 1839, American Museum of Natural History, New York; see also entry for January 1, 1839, pp. 98–99.

28. For an extended discussion, see Daniel C. Haskell, *The United States Exploring Expedition, 1828–1842, and Its Publications, 1844–1874* (New York: New York Public Library, 1942), pp. 1–28.

29. By an Act of Congress, August 26, 1842, quoted in ibid., p. 150.

30. Ibid., pp. 54–61.

31. Titian R. Peale, "U.S. Exploring Expedition," typescript, Titian R. Peale Papers, American Museum of Natural History, New York.

32. Titian R. Peale to Thaddeus W. Harris, January 12, 1854, Museum of Comparative Zoology, Harvard University.

33. George Ord to Charles Waterton, November 21, 1842, typescript from the original owned by R. Milnes Walker, American Philosophical Society.

34. George Ord to Charles Waterton, May 24, 1846, microfilm, American Philosophical Society.

35. This distinction, articulated by Charles Pickering to Samuel G. Morton, January 25, 1838, is quoted by Richard G. Beidelman, "Some Biographical Sidelights on Thomas Nuttall, 1786–1859," *Proceedings of the American Philosophical Society* 104 (1960):95–96, from the Samuel G. Morton Papers, American Philosophical Society, Philadelphia.

Chapter 11: The Business of Science

1. William Maclure, *Opinions on Various Subjects,* 3 vols. (New Harmony: School Press, 1831–38), 3:224.

2. For an extended discussion, see William H. Goetzmann, *Exploration and Empire: The Explorer and the Scientist in the Winning of the American West* (New York: Alfred A. Knopf, 1971), pp. 57–64, 182–84.

3. See Roger L. Nichols, "Stephen Long and Scientific Explorations on the Plains," *Nebraska History* 52 (Spring 1971):51–64.

4. According to Ralph Schwartz, director of New Harmony, Inc. (personal communication), in August 1825 these naturalists persuaded Maclure to visit New Harmony to convince him of the merits of Owen's venture.

5. C. S. Rafinesque, *A Life of Travels* (Philadelphia: Printed for the author, 1836), pp. 78–79: "I had some intention to join Mr. Maclure at New Harmony, but he had friends jealous of me also: it was well for me since his views and fine Colleges have been abortive."

6. A. O. Weese, ed., "The Journal of Titian Ramsay Peale, Pioneer Naturalist," *Missouri Historical Review* 41 (1947):156, 266–67, 271.

7. For more on the troubles that beset the transport of the expedition's scientific party, see Charlotte M. Porter, "The American West Described in Natural History Journals, 1819–1836," *Midwest Review* 2 (Spring 1980):25–26.

8. The cause of scurvy was not understood, and the sick were "confined principally to the woods to subsist on game." See Roger L. Nichols, ed., *The Missouri Expedition (1818–1820)* (Norman: University of Oklahoma Press, 1969), p. 83, a journal account presumably written by John Gale, the expedition's surgeon. This shocking situation was confirmed by Edwin James, *Account of an Expedition from Pittsburgh to the Rocky Mountains Performed in the Years 1819, 1820,* 2 vols. (Philadelphia: H. C. Carey and I. Lea, 1823), 1:195.

9. Rafinesque continued to advocate Maclure's ideas; see his *The Pleasures and Duties of Wealth* (Philadelphia: Eleutherium of Knowledge, 1840), pp. 11–31.

10. Jessie Poesch, *Titian Ramsay Peale and His Journals of the Wilkes Expedition* (Philadelphia: American Philosophical Society, 1961), p. 113.

11. See Clark A. Elliott, *Biographical Dictionary of American Science* (Westport, Conn.: Greenwood Press, 1979), pp. 251–52.

12. L. H. Pammell, "Dr. Edwin James," *Annals of Iowa* 3d ser., 8 (October 1907):294.

13. George Ord to Charles Waterton, August 11, 1850, microfilm, American Philosophical Society, Philadelphia.

14. C. S. Rafinesque, *Autikon Botanikon* (Philadelphia: 1815, 1840 [probably actually printed in 1838]), p. 3.

15. Gray and Torrey monitored the initial selection of the scientific corps. DeKay and Gray were first approached, but declined and arranged instead for the appointments of James Dwight Dana and Charles Pickering. John Witt Randall (the entomologist) had only recently graduated from Harvard, where Horatio Hale (the philologist) was still a student. See William Stanton, *The Great United States Exploring Expedition of 1838–1842* (Berkeley and Los Angeles: University of California Press, 1975), pp. 43–48.

16. Daniel C. Haskell, *The United States Exploring Expedition, 1838–1842, and Its Publications, 1844–1874* (New York: New York Public Library, 1942), p. 17.

17. Stanley M. Guralnick, *Science and the Ante-Bellum American College* (Philadelphia: American Philosophical Society, 1975), pp. 109–11.

18. Harry B. Weiss and Grace M. Ziegler, *Thomas Say, Early American Naturalist* (Baltimore: Charles C. Thomas, 1931), p. 125.

19. C. S. Rafinesque to John Torrey, September 24, 1838, Boston Public Library,

typescript kindly provided by Charles Boewe, editor of the Papers of C. S. Rafinesque.

20. John K. Townsend, *Narrative of a Journey across the Rocky Mountains to the Columbia River and a Visit to the Sandwich Islands, Chili &c.* (Philadelphia: Henry Perkins, 1839), in Reuben Gold Thwaites, ed., *Early Western Travels,* vol. 21 (Cleveland: Arthur H. Clark, 1905), p. 325. Thornburg, the fort tailor, was later killed by Hubbard, the gunsmith, on another drunken spree.

21. H. M. Fuller and L. R. Hafen, eds., *The Journal of Captain John R. Bell* (Glendale, Calif.: Arthur H. Clark, 1957), p. 165.

22. Thomas Nuttall to Asa Gray, April 23, 1841, Historic Letter File, Gray Herbarium, Harvard University.

23. See also Thomas Nuttall to Asa Gray, March 29, 1841, ibid.

24. C. S. Rafinesque to John Torrey, June 9, 1819, Torrey Papers, New York Botanical Garden, Bronx, New York; C. S. Rafinesque to John Torrey, June 9, 1829, Yale University; microfilm, T. J. Fitzpatrick Collection, Spencer Library, University of Kansas, Lawrence; see also T. J. Fitzpatrick, *Rafinesque: A Sketch of His Life with Bibliography* (Des Moines: Historical Department of Iowa, 1911); R. E. Call, *The Life and Writings of Rafinesque* (Louisville: John P. Morton, 1895); and Elmer D. Merrill, "A Life of Travels by C. S. Rafinesque," *Chronica Botanica* 8 (1944):292–360, a reprint of Rafinesque's autobiography of the same title published in Philadelphia in 1836.

25. C. S. Rafinesque to Thomas Jefferson, January 25, 1821, in Edwin M. Betts, "The Correspondence between Constantine Samuel Rafinesque and Thomas Jefferson," *Proceedings of the American Philosophical Society* 87 (1944):372–73.

26. See the rare prospectus for Peale's elaborate *Lepidoptera Americana: or Original Figures of the Moths and Butterflies of North America: in their Various Stages of Existence and the Plants on which they Feed* (Philadelphia: William P. Gibbons, 1833), at the American Museum of Natural History.

27. Titian R. Peale to Thaddeus Harris, January 12, 1854, Miscellaneous Correspondence, Museum of Comparative Zoology, Harvard University.

28. Susan Sheets-Pyenson, "War and Peace in Natural History Publications: The Naturalist's Library, 1833–1842," *Isis* 72 (March 1981):52.

29. Robert Jameson, ed., *The American Ornithology,* vols. 71–74 in *Constable's Miscellany of Original and Selected Publications in the Various Departments of Literature, Science, and the Arts* (Edinburgh: Constable & Co.; London: Hurst, Chance & Co., 1831), claimed that the work was "regularly arranged for the first time" by orders. Numerous favorable reviews of both this and Jardine's edition are included in the Wilson Papers, Houghton Library, Harvard University.

30. Sheets-Pyenson, "War and Peace," p. 63.

31. See Lucy Say to Thaddeus Harris, January 24, 1835, and Thomas Say to Thaddeus Harris, May 20, November 28, 1830, and April 1834, Harris Papers, Houghton Library, Harvard University.

32. See William Cooper to Charles Lucien Bonaparte, September 26, 1827, microfilm, American Philosophical Society.

33. William Cooper to Charles Lucien Bonaparte, November 22, 1830, microfilm, American Philosophical Society.

34. William Cooper to Charles Lucien Bonaparte, October 14, 1831, microfilm, American Philosophical Society.
35. Cooper, who also aided Torrey, Nuttall, and Audubon, claimed to have tried everything, including tobacco leaves, to ward off destructive insects; see his letter to C. L. Bonaparte, June 27, 1831, microfilm, American Philosophical Society.
36. H.E.U. to Isaac Hays, December 8, 1826, Hays Papers.
37. Antipathy stemming from a malpractice suit may have motivated Godman to try to outdo Harlan by publication of his more elaborate book on mammals; see H.E.U. to Isaac Hays, Hays Papers, American Philosophical Society; and S. D. Gross, *Autobiography of Samuel D. Gross, M.D.* (Philadelphia: George Barrie, 1887), p. 45.
38. See, for example, John Godman, *American Natural History,* 3 vols. (Philadelphia: H. C. Carey and I. Lea, 1826–28), 1:vii.
39. This curious fact is extensively discussed by Charlotte M. Porter, "The Lifework of Titian Ramsay Peale," *Proceedings of the American Philosophical Society* 129 (1985):306–08.
40. John Godman married Angelica K. Peale, Rembrandt Peale's daughter, on October 6, 1821. For his friendship with Titian Peale, see his *American Natural History,* 1:58.
41. See Frank L. Burns, "Miss Lawson's Recollections of Ornithologists," *Auk* 34 (July 1917):280.
42. The plates in Godman, *American Natural History,* vol. 3, after page 242, are perhaps the last known work from the old master's hand.
43. H.E.U. to Isaac Hays, August 17, June 17, 1827, Hays Papers.
44. H.E.U. to Isaac Hays, August 4, 1827, Hays Papers.
45. Burns, "Miss Lawson's Recollections," p. 276.
46. Francis Hobart Herrick, *Audubon the Naturalist,* 2 vols. (New York: D. Appleton, 1917), 1:328–34.
47. See H.E.U. to Isaac Hays, January 26, February 1, 1827, Hays Papers.
48. Translator's note in Bernhardt, Duke of Saxe-Weimar Eisenach, *Travels through North America during the Years 1825 and 1826,* 2 vols. (Philadelphia: Carey, Lea & Carey), 2:108.
49. Details are given by Lucy Say to Thaddeus Harris, January 24, 1835, Harris Papers.
50. Benjamin H. Coates, *Biographical Sketch of the Late Thomas Say, Esq.* (Philadelphia: Published by the Academy, 1835).
51. Emma Jones Lapsansky, " 'Since They Got Those Separate Churches': Afro-Americans and Racism in Jacksonian Philadelphia," *American Quarterly* 32 (Spring 1980):54.
52. Samuel N. Rhoads, "George Ord," *Cassina,* no. 12, (1908), p. 4.
53. Robert Henry Welker, *Birds and Men: American Birds in Science, Art, Literature, and Conservation, 1800–1900* (New York: Atheneum, 1966), pp. 50–51; George Ord to Charles Waterton, April 23, 1832, microfilm, American Philosophical Society.
54. Ord charged Audubon with plagiarism (specifically copying Wilson's Mississippi kite) at an August meeting recorded in the *Proceedings of the American Philosophical Society,* n.s., 1 (1840):273–74.

55. George Ord to Charles Waterton, November 21, 1842, typescript from the original, American Philosophical Society.

56. See George Ord, "A Memoir of Charles-Alexandre LeSueur," *American Journal of Science and Arts*, 2d ser., 8 (1849):189–216.

57. Thomas Say to Thaddeus W. Harris, August 7, 1834, Harris Papers.

58. Thomas Say to Samuel G. Morton, March 5, 1833, Morton Papers, American Philosophical Society, quoted by Nathan Reingold, ed., *Science in Nineteenth Century America: A Documentary History* (New York: Hill & Wang, 1964), pp. 39–40.

59. Asa Gray, "The Longevity of Trees," *North American Review* 59 (1844):191.

60. Ord, "Memoir," pp. xiii–xiv.

61. Lucy Say to Thaddeus Harris, January 6, 1835, Harris Papers.

62. Quoted by Arnold Mallis, *American Entomologists* (New Brunswick: Rutgers University Press, 1971), pp. 22–23; Lucy Say to Thaddeus Harris, March 25, 1835, Museum of Comparative Zoology.

63. Waldemar H. Fries, *The Double Elephant Folio* (Chicago: American Library Association, 1973), p. 7. For more on their friendship see Alexander B. Adams, *John James Audubon* (New York: G. P. Putnam's Sons, 1966), pp. 261–64, and numerous references in Alice Ford, *John James Audubon* (Norman: University of Oklahoma Press, 1964).

64. See the author's preface to Alexander Philip Maximilian, Prince of Wied-Neuwied, *Travels in the Interior of North America, 1832–1834,* in Reuben Gold Thwaites, ed., *Early Western Travels* (Cleveland: Arthur H. Clarke, 1906), 22:29–30.

65. First appearing in the third volume of the *Journal of the Academy of Natural Sciences* in 1824 and continuing through volume 5 of the next year, this supplement of more than two hundred pages was completed in the second volume of the New-York Lyceum's *Annals of Natural History.*

66. For a wonderful discussion of this subject, see R. J. Cooter, "Phrenology: The Provocation of Progress," *History of Science* 14 (1976):211, 213, 226–27.

67. Ord, "Memoir," pp. vii–ix. See also Steven Shapin, "The Pioneer Days of Progress," *Times Literary Supplement,* January 7, 1977, p. 3.

68. Amos Eaton to John Torrey, July 21, 1817, Torrey Papers.

69. S. S. Haldeman, "Notice of the Zoological Writings of the Late C. S. Rafinesque," *American Journal of Science* 42 (1842):286.

70. Asa Gray, "Notice of the Botanical Writings of the Late C. S. Rafinesque," *American Journal of Science* 40 (January–March 1841):230–31, 234, 239.

71. In 1835, Rafinesque claimed to have worked out the correspondences between "Endogens" (dicots) and vertebrates and between "Exogens" (monocots) and "unboney Animals"; see his *Flora Telluriana* (Philadelphia: Printed for the author by H. Probasco, 1836), pp. 46–48.

72. Many were published in *American Monthly Magazine* and *Critical Review* 1 (October 1817):426–30; 2 (March 1818):342–44; 3 (June 1818):96–101; 4 (August 1818):264–74; see also Charles Boewe, ed., Fitzpatrick, *Bibliography,* nos., 265, 267, 268; C. S. Rafinesque to Amos Eaton, March 15, 1818, T. J. Fitzpatrick Collection, Spencer Library, University of Kansas, Lawrence; and Amos Eaton to John Torrey, October 5, 1817, Torrey Papers.

73. William Baldwin to William Darlington, February 9, 1819, in William

Darlington, ed., *Reliquiae Baldwinianae* (Philadelphia: Kimber and Sharpless, 1843), p. 300. See also editor's note 2 in Gray, "Notice," p. 237.

74. C. S. Rafinesque to John Torrey, June 9, 1819, Torrey Papers.

75. James DeKay to C. L. Bonaparte, January 29, 1825, microfilm, American Philosophical Society.

76. Say's comments were relayed by DeKay to Bonaparte, ibid.

77. See Charles Boewe, "Editing Rafinesque Holographs: The Case of the Short Letters," *Filson Club Historical Quarterly* 54 (January 1980):41.

78. Thomas Say, "Mammalia," MS notes, Museum of Comparative Zoology.

79. Asa Gray to Benjamin Silliman, October 5, 1840, microfilm, American Philosophical Society.

80. Asa Gray to Benjamin Silliman, November 15, 1840, microfilm, American Philosophical Society.

81. Asa Gray to Benjamin Silliman, December 31, 1840, microfilm, American Philosophical Society.

82. See Rafinesque, *Flora Telluriana*, p. 25; also Boewe, "Editing Rafinesque Holographs," pp. 39–40; Merrill, "Life of Travels," p. 341; Wayne Hanley, *Natural History in America* (New York: Quadrangle, New York Times Book Co., 1977), p. 137; Elmer D. Merrill, "Rafinesque's Publications from the Standpoint of World Botany," *Proceedings of the American Philosophical Society* 87 (1943):110–19; Merrill, "A Generally Overlooked Rafinesque Paper," *Proceedings of the American Philosophical Society* 86 (1942):72–90.

83. David Starr Jordan, "Rafinesque," *Popular Science Monthly* 29 (June 1886):218.

Chapter 12: Science by Management

1. Richard Harlan to John James Audubon, November 19, 1828, Miscellaneous Manuscripts, Harlan, New-York Historical Society, New York.

2. See C. S. Rafinesque to John Torrey, April 12, 1828, February 10, 1834, Torrey Papers, New York Botanical Garden, Bronx, New York; and C. S. Rafinesque to John Torrey, October 1, 1839, microfilm, T. J. Fitzpatrick Collection, Spencer Library, University of Kansas, Lawrence.

3. Howard Mumford Jones, *America and French Culture, 1750–1848* (Chapel Hill: University of North Carolina Press, 1927), p. 154.

4. Robert E. Bieder, "Albert Gallatin and the Survival of Enlightenment Thought in Nineteenth-Century American Anthropology," in Timothy H. H. Thoreson, ed., *Toward a Science of Man: Essays in the History of Anthropology* (The Hague: Mouton, 1975), pp. 94–95; Courtney Robert Hall, *A Scientist in the Early Republic* (New York: Columbia University Press, 1934), p. 115; Frank Spencer, "Two Unpublished Essays on the Anthropology of North America by Benjamin Smith Barton," *Isis* 68 (December 1977):569; Edward Lurie, "Louis Agassiz and the Races of Man," in Nathan Reingold, ed., *Science in America since 1820* (New York: Science History Publications, 1976), pp. 148–54.

5. C. S. Rafinesque, "Complexions of Mankind," *Atlantic Journal and Friend of Knowledge* 1 (1832–33):172–73, a rare periodical available as a photolithograph (Cambridge, Mass.: Murray, 1946).

6. Notice for the *Troy Whig*, January 20, 1835, quoted in Ethel M. McAllister,

Amos Eaton, Scientist and Educator, 1776–1842 (Philadelphia: University of Pennsylvania Press, 1941), p. 247.

7. See Charles Pickering, *Races of Man: And Their Geographical Distribution* (Boston: Charles C. Little & James Brown, 1848), pp. 302ff. Pickering's presentation copy to Morton can be seen at the Academy of Natural Sciences. See also W. W. Ruschenberger's posthumous preface to Charles Pickering's, *Chronological History of Plants: Man's Record of His Own Existence Illustrated through Their Names, Uses, and Companionship* (Boston: Little, Brown, 1879), p. xiv, citing Gray's laudatory obituary of Pickering in the *Proceedings of the American Academy of Arts and Sciences.*

8. See Reginald Horsman, "Scientific Racism and the American Indian in the Mid-Nineteenth Century," *American Quarterly* 27 (1975):153; also Hans Aarsleff, "An Outline of Language-Origins Theory since the Renaissance," *Annals of the New York Academy of Science* 280 (1976):9, 11, 15; see also R. Reese, ed., "List of Books and Pamphlets in the Library of the Workingman's Institute, New Harmony, Indiana," New Harmony, March 1909, mimeograph, Widener Library, Harvard University.

9. Thomas Nuttall, *A Popular Handbook of the Ornithology of Eastern North America,* ed. Montague Chamberlain, 2d rev. ed., 2 vols. (Boston: Little, Brown, 1896), 1:22.

10. C. S. Rafinesque, "Botanical Geography," in his *New Flora and Botany of North America* (1836; rpt. Cambridge, Mass.: Murray, 1946), p. 31.

11. Richard Harlan, "On the Successive Formations of Organized Beings," in his *Medical and Physical Researches* (Philadelphia: Lydia R. Bailey, 1835), p. 240.

12. Very little is known about Rafinesque's training or research before his arrival in the United States, but see L. B. Holthuis and M. Boeseman, "Notes of C. S. Rafinesque-Schmaltz's (1810) Caratteri di alcuni nuovi generi e nuove specie di animali e pianti della Sicilia," *Journal of the Society for the Bibliography of Natural History* 8 (1977):231–34.

13. [Robert Chambers], *Vestiges of the Natural History of Creation,* vol. 2 (London: John Churchill, 1844), plate 25, showing the enlarged parts of *Mantispa.*

14. Quoted in A. Hunter Dupree, *Asa Gray* (Cambridge, Mass.: Belknap Press of Harvard University Press, 1959), p. 146.

15. [Asa Gray], "The Longevity of Trees," *North American Review* 59 (1844):193.

16. Thomas Nuttall, *The North American Sylva of F. Andrew Michaux,* 3 vols. (Philadelphia: J. Dobson, 1842–52), 3:91.

17. Quoted by G. B. Goode, "The Beginnings of Natural History in America," *Annual Report of the Smithsonian Institution, National Museum,* pt. 2 (Washington, D.C.: U.S. Government Printing Office, 1901), p. 405.

18. John Bartram's report was appended to William Stork's *Account of East Florida* (1766) to "puff off" the first edition, which had resulted in 120 applications for land grants to the British Board of Trade; see Edmund Berkeley and Dorothy Smith Berkeley, *The Life and Travels of John Bartram* (Tallahassee: University Presses of Florida, 1982), pp. 267–71.

19. See John Lindley, *The Vegetable Kingdom,* 3d ed. (London: Bradbury & Evans, 1853), p. xii.

20. Dupree, *Gray,* pp. 211–12.

21. Theodore Dwight Bozeman, *Protestants in an Age of Science: The Baconian*

Ideal and Antebellum American Religious Thought (Chapel Hill: University of North Carolina Press, 1977), pp. 167–77.

22. See the photograph of Gray in the field, 1877, Historic Letter File, Gray Herbarium, Harvard University.

23. Peale's asking price was a "very low" $550 according to C. V. Riley to Spencer Baird, November 15, 1884, and C. V. Riley's response to T. R. Peale dated November 1 [1884?], Record Unit 139, Smithsonian Archives, Washington, D.C.

24. See Robert H. Walker, *The Poet and the Gilded Age: Social Themes in Late Nineteenth Century American Verse* (Philadelphia: University of Pennsylvania Press, 1963), pp. 14–35, especially the discussion of utopia on p. 16; for a general introduction, see George Daniels, ed., *Darwinism Comes to America* (Waltham: Blaisdell, 1968), pp. 1–28.

25. William Swainson, *A Preliminary Discourse on the Study of Natural History* (London: Longman, Rees, Orme, Brown, Green & Longman, 1834), p. 3. Swainson made a living prefacing such inexpensive tracts as this and other British pocketbooks developed along the lines of Constable's *Miscellany*.

26. Nathan Reingold, "American Indifference to Basic Research: A Reappraisal," in George Daniels, ed., *Nineteenth Century American Science: A Reappraisal* (Evanston: Northwestern University Press, 1972), p. 43.

27. Bernard Houghton, *Scientific Periodicals: Their Historical Development, Characteristics, and Control* (Hamden, Conn.: Linnet Books, 1975), p. 27.

28. J. E. Smith, "Early Invertebrate Zoology: Men and Their Animals," in S. Zuckerman, ed., *The Zoological Society of London 1826–1976 and Beyond* (London: Academic Press, 1976), pp. 73–74.

Chapter 13: The Peaceable Kingdom

1. William Maclure, *Opinions on Various Subjects,* 3 vols. (New Harmony: School Press, 1831–38), 1:6–7, 52–53.

2. Benjamin Smith Barton, *Elements of Botany,* 2 vols. (Philadelphia: Printed for the author, 1803); see also Edmund Berkeley and Dorothy Smith Berkeley, *The Life and Travels of John Bartram* (Tallahassee: University Presses of Florida, 1982), p. 228; Raymond Phineas Stearns, *Science in the British Colonies of America* (Urbana: University of Illinois Press, 1970), pp. 573–619.

3. Ann Leighton, *American Gardens in the Eighteenth Century* (Boston: Houghton Mifflin, 1976), pp. 43–46.

4. William Bartram, *Travels through North & South Carolina, Georgia, East & West Florida . . .* (Philadelphia: James & Johnson, 1791), p. 155.

5. John Prest, *The Garden of Eden* (New Haven: Yale University Press, 1981), pp. 102–7.

6. See J. Percy Moore, "William Maclure, Scientist and Humanitarian," *Proceedings of the American Philosophical Society* 91 (1947):234–49.

7. Maclure's popular standing is evident in the unsigned "Literary Intelligence," *Port Folio,* 4th ser., 3 (November 1819):434, which compares American and British intellectual prowess, using Wilson and Maclure, both of whom were born in Scotland.

8. David Tatham, "Edward Hicks, Elias Hicks and John Comly: Perspectives on

the Peaceable Kingdom Theme," *American Art Journal* 13 (Spring 1981):47–48.

9. Jane Kallir, *The Folk Art Tradition: Naive Painting in Europe and the United States* (New York: Viking Press, 1981), pp. 31–32, exhibition catalog for the Galerie St. Etienne; see also Kallir's essay for the checklist, "The Folk Art Tradition," November 17, 1981, through January 9, 1982, Galerie St. Etienne, New York.

10. Alice Ford states in her definitive biography, *Edward Hicks: Painter of the Peaceable Kingdom* (Philadelphia: University of Pennsylvania Press, 1952), p. 119, that there is an invitation to a traveling menagerie in the Hicks Collection, Newtown, Pennsylvania.

11. For example, see the chapter entitled "The Curious Quaker," in Joseph Kastner, *A Species of Eternity* (New York: Alfred A. Knopf, 1977), pp. 40–67. See also William Bartram, *Botanical and Zoological Drawings,* ed. and intro. Joseph Ewan (Philadelphia: American Philosophical Society, 1968).

12. Leighton, *American Gardens,* pp. 43–46.

13. Ford, *Edward Hicks,* pp. 33, 118. The first recognized painting of Niagara Falls, Vanderlyn's 1802 canvas, now at the Albany Institute of History and Art, was engraved by Merigot in 1804.

14. Wilson's American popularity is reviewed by Hans Huth, *Nature and the American* (Lincoln: Bison Book, 1972), pp. 25–26. A contemporary writer was "struck with the fact that his prose is poetical, while the poetry inclines to the prosaic." See William B. O. Peabody, "Life of Alexander Wilson," in Jared Sparks, comp., *The Library of American Biography,* 25 vols. (Boston: Hilliard, Gray, 1834–48), 2:91.

15. Alexander Wilson to Charles Orr, July 23, 1800, Letters from the Library of Sir William Jardine, Houghton Library, Harvard University.

16. Published serially through March 1810, Alexander Wilson's "The Foresters, Description of a Pedestrian Tour to the Fall of Niagara in the Autumn of 1804," first appeared in the *Port Folio* 2 (July 1809):70–77. G. Cooke's engraving after a drawing by Wilson accompanied the second part issued in August 1809, opposite page 147. Wilson's poem "A Pilgrim" appeared with narrative in the *Port Folio* 3 (June 1810):499–511.

17. John L. Gardiner to Alexander Wilson, April 30, 1810, in John H. Sage, "An Historic Letter," *Auk* 7 (October 1895):360.

18. Frederick B. Tolles, "The Primitive Painter as Poet," *Bulletin of the Friends Historical Association* 50 (Spring 1969):12–30; see Hicks's couplets in the lettered border to his *Peaceable Kingdom,* ca. 1840, Friends Historical Library of Swarthmore College, Swarthmore, Pennsylvania.

19. Ford, *Edward Hicks,* p. 33. The New-York Historical Society owns a copy of this special printing, funded by the poet's brother, Joseph Wilson, and S. Siegfried and the first of a series of local printings.

20. Two good essays in exhibition catalogs chronicle Hicks's progress as a painter: [Leon A. Arkus], "Edward Hicks, 1780–1849," in *Hicks, Kane, Pippin* (Pittsburgh: Museum of Art, Carnegie Institute, 1966), unpaginated; and Eleanor Price Mather, "Edward Hicks," in Jean Lipman and Tom Armstrong, eds., *American Folk Painters of Three Centuries* (New York: Hudson Hills Press, with the Whitney Museum of American Art, 1980), pp. 88–97.

21. The painting belonged to Colonel George Bomford, whose first wife's father was the artist Charles Catton (1756–1819), a heraldic painter who settled in Ulster County, New York. Peale borrowed the picture in May 1819, deposited it in the museum on June 7, 1819, began copying it in October, and finished his version with the "aid of high magnifying powers of spectacles" by December 11. Whether the original was on view from June to October is unclear. See "Memoranda of the Philadelphia Museum," Historical Society of Pennsylvania, and Peale's letters to his sons Rembrandt Peale, December 16, 1819, and Titian Ramsay Peale, December 25, 1819, Charles Willson Peale Papers, American Philosophical Society, Philadelphia, quoted by Charles Coleman Sellers, "Portraits and Miniatures by Charles Willson Peale," *Transactions of the American Philosophical Society* 42 (June 1952):44. Ford, *Edward Hicks*, p. 23, suggests another view.

22. Mary Bartlett Cowdrey, *American Academy of Fine Arts and American Art-Union Exhibition Record, 1816–1852* (New York: New-York Historical Society, 1953), p. 61. Another landscape painting lent by M. E. Hicks was exhibited in 1820, although there is no record of an M. E. Hicks in the Hicks Family File, New-York Historical Society. According to William Dunlap, *History of the Rise and Progress of the Arts of Design in the United States,* 2 vols. (1834; rpt. New York: Dover, 1969), vol. 2, pt. 1, p. 210, Samuel Maverick, the son of the engraver Peter Maverick, owned two genre pictures by Catton.

23. Wilson's original, a pencil drawing of the bald eagle at Niagara Falls, can be viewed at the Department of Ornithology, Museum of Comparative Zoology, Harvard University.

24. This "White-Headed Eagle," as Peale preferred to call it, was a museum pet that had lived for years in a cage high above the menagerie Peale maintained.

25. The impression at the New-York Historical Society is too large to have been a frontispiece for the 1818 reprint of Wilson's "The Foresters" but could have been intended to accompany the passage in the poem describing Niagara Falls, *Port Folio* 2 (September 1809):273–78.

26. Frank L. Burns, "Miss Lawson's Recollections of Ornithologists," *Auk* 34 (July 1917):277. Barralet's profile was engraved for the *Port Folio* and later editions of Wilson's work; see also Dunlap, *History,* vol. 2, pt. 1, p. 43.

27. Robert H. Welker, *Birds and Men: American Birds in Science, Art, Literature, and Conservation, 1800–1900* (New York: Atheneum, 1966), p. 53.

28. These are *The Great Horseshoe Falls, Niagara* (1820) and *A General View of the Falls of Niagara* (1820), both at the National Museum of American Art, Smithsonian Institution. There are at least two other versions of about the same date as well. Fisher's work was frequently engraved for almanacs and keepsake books from 1825 to 1865.

29. Hicks was a prodigious worker. To date, sixty paintings of his *Peaceable Kingdom* are known. In the spring of 1980, two sold at auction for $210,000 and $270,000, and a third fetched $450,000 at a private sale.

30. Ford, *Edward Hicks*, p. 16.

31. "The Foresters" was originally published with footnotes that included the scientific nomenclature for the bald eagle. See also Peabody, "Life of Wilson," p. 92, and Alexander B. Grosart, *The Poems and Literary Prose of Alexander Wilson,* 2 vols. (Paisley, Scotland: Alex. Gardener, 1876), 1:111–13.

32. James E. Ayres, "Edward Hicks and His Sources," *Antiques* (February 1976):366–68.
33. For more on the popularity and use of geographical illustrations during the 1820s, see Maybelle Mann, "American Landscape Prints," *Arts & Antiques* 4 (May–June 1981):90–97.
34. William Maclure, *Observations on the Geology of the United States of America* (Philadelphia: Printed for the author by Abraham Small, 1817).
35. Jessie Poesch, *Titian Ramsay Peale and His Journals of the Wilkes Expedition* (Philadelphia: American Philosophical Society, 1961), p. 60; Walter W. Ristow, "Lithography and Maps, 1796–1850," in David Woodward, ed., *Five Centuries of Map Printing* (Chicago: University of Chicago Press, 1975), p. 79.
36. Dunlap, *History*, vol. 2, pt. 1, p. 212.
37. Kearny also engraved plates for the British edition of Edwin James, *Account of an Expedition from Pittsburgh to the Rocky Mountains, Performed in the Years 1819, 1820*, 2 vols. (London: Longman, Hurst, Rees, Orme & Brown, 1823), which was based in part on Titian Peale's notes, drawings, and specimens.
38. The *Port Folio*, n.s., 1 (June 1809), includes an example of Barralet's work, a drawing of the Pennsylvania Academy of Fine Arts (facing p. 461). Tiebout engraved plates for Thomas Say's 1824 and 1825 volumes.
39. The ticket is reproduced in William E. Wilson, *The Angel and the Serpent* (Bloomington: University of Indiana Press, 1984), after p. 146.
40. For this information, I am grateful to Ralph G. Schwarz, president of Historic New Harmony, Inc.
41. Alexander Philip Maximilian, Prince of Wied-Neuwied, *Travels in the Interior of North America, 1832–1834*, trans. H. Evans Lloyd, 2 vols. (London: Ackermann 1843), 1:74–92, describes the community and its library.
42. First published in German in 1839, Maximilian's popular *Travels* was issued in French and English editions, all of which included a picture atlas with eighty-one colored plates after drawings by Karl Bodmer.
43. Earlier in his career, Tiebout had engraved *A Survey of the Roads of the United States of America, 1789–1792* for Christopher Colles. Henry S. Tanner pursued a similar course with his *Description of the Canals and Rail Roads of the United States* in 1840 and *The Traveller's Hand Book for the State of New York and the Province of Canada . . . with Maps* in 1844, both published in New York by T. R. Tanner.
44. Ayres, "Edward Hicks," p. 367, compares the figure of the well-known Quaker minister in *The Residence of the Late Richard Jordan* (a transfer-printed pattern, ca. 1836, produced by Joseph Heath and Company of Tunstall, Staffordshire, from an engraving by Kearny after a drawing by W. Mason) to Hicks's figure of William Penn in *Penn's Treaty*, ca. 1830, Abby Aldrich Rockefeller Folk Art Center, Williamsburg, Virginia.
45. Hicks was apparently ignorant of the hard facts of Penn's "Holy Experiment." Penn referred to Indian rights as an "incumbrance," which he pledged to eliminate so as to guarantee good title to his patentees; see Francis Jennings, *The Invasion of America: Indians, Colonialism, and the Cant of Conquest* (Chapel Hill: University of North Carolina Press, 1975), p. 130n.
46. Alexander Wilson to William Bartram, December 24, 1804, Wilson Papers, Houghton Library, Harvard University. For popular views, see William Bar-

ham, *Descriptions of Niagara, Selected from Various Travellers with Original Additions* (Gravesend: Published by the compiler, n.d. [but before 1850]), particularly the account of Mrs. L. H. Sigourney, pp. 111–17. For scientific views, see Samuel Latham Mitchill, "The Original Saltines of the North American Lakes," in his *Observations on the Geology of North America* (New York: Kirk and Mercein, 1818), pp. 352–58, and Chester Dewey, "Configuration of the County Lying South of Lake Erie," ibid., p. 419.

47. See, for example, the editor's note to "The Natural Bridge," a poem by John Davis in the *Port Folio* 7 (January 1809):82–84.

48. See the long unsigned "Notice and Review of the 'Reliquiae Diluviae,' or Observations on the Organic Remains Contained in Caves, Fissures and Diluvial Gravel, and on Other Geological Phenomena Attesting the Action of an Universal Deluge," *American Journal of Science* 8 (August 1824):150–68, and "The New Theory of the Earth," *North American Review* 28 (1829):265–66, in which the author, Edward Hitchcock, a theology student turned geologist, advocated "a fastidious avoidance of hypothesis."

49. Jefferson's famous description of the Natural Bridge is found in "Query Six" of *Notes on the State of Virginia* (Philadelphia: R. T. Rawle, 1801), pp. 44–45.

50. Francis W. Gilmer, "On the Geological Formation of the Natural Bridge of Virginia," *Transactions of the American Philosophical Society*, n.s., 1, (1818):189–90. Gilmer was a lawyer who aided Jefferson in starting the University of Virginia.

51. Edward Hicks to John Watson, December 7, 1829, March 16, 1839, Galerie St. Etienne, New York.

52. Thomas Hicks, *Portrait of Edward Hicks* (1838), oil on canvas, Abby Aldrich Rockefeller Folk Art Center, Williamsburg. See Benjamin Rush, *Medical Inquiries and Observations upon the Diseases of the Mind* (Philadelphia: Kimber & Richardson, 1812), pp. 343, 122, the earliest work on psychiatry by an American.

53. Edward Hicks, *Memoirs of the Life and Religious Labors of Edward Hicks, Late of Newtown, Bucks County, Pennsylvania Written by Himself* (Philadelphia: Merrihew & Thompson, 1851), p. 330. Because Hicks did not begin his retrospective autobiography until 1843, many early influences and events are unrecorded; see Alice Ford, "The Publication of Edward Hicks' Memoirs," *Bulletin of the Friends Historical Association* 50 (Spring 1961):5.

54. Peale's original drawing and lithograph, which differ slightly, can be compared at the American Philosophical Society. LeSueur's illustration appears to be the source for the grizzly bear crudely depicted in *Davy Crockett's Almanack* for 1837, certainly a popular source, but one available to Hicks only after he had begun to paint the upright bear ca. 1833.

55. Edward Hicks, *Washington Crossed Here* (ca. 1848–49), oil on canvas, private collection.

56. John Godman, *American Natural History*, 3 vols. (Philadelphia: H. C. Carey and I. Lea, 1826), 1:179.

57. Poesch, *Peale*, p. 44.

58. Godman, *American Natural History*, 1:185–86.

59. Charles Coleman Sellers, *Mr. Peale's Museum* (New York: W. W. Norton, 1980), p. 315, as listed in the sheriff's sale catalog of 1848.

60. Tatham, "Edward Hicks," p. 48.
61. See Sandra Brand and Elissa Cullman, *Small Folk: A Celebration of Childhood in America* (New York: E. P. Dutton with the Museum of Folk Art, 1980), pp. 8, 29, and illustrations of works by Ammi Phillips, William Bartoll, Henry Walton, Erastus Salisbury Field, and Joseph Whiting Stock. Of course, Hicks's approach differed from their attention to real pets and real playthings, although he, too, may have been influenced by the Transcendental view of children as one with nature.
62. For examples, see Charles Willson Peale, *A Scientific and Descriptive Catalogue of Peale's Museum* (Philadelphia: Samuel H. Smith, 1796), pp. 27–28; C. W. Peale to Rubens Peale, August 19, 1805, quoted by Poesch, *Peale,* p. 11, from Letterbook Six, p. 139, Charles Willson Peale Papers, American Philosophical Society.
63. Charles Willson Peale's "Autobiography," typescript, p. 219, Charles Willson Peale Papers.
64. Charles Willson Peale, *The Artist's Mother, Mrs. Charles Peale, with Her Grandchildren* (ca. 1783), oil on canvas, private collection.
65. See "Original Letter from Sir Benjamin West to Charles W. Peale, Esq.," *Port Folio* 3 (July 1810):8.
66. G.M., "The Exhibition," *Port Folio,* n.s., 8 (August 1812):149–50. For Maclure's popular standing see the unsigned "Literary Intelligence," *Port Folio,* 4th ser., 3 (November 1819):434, which compares American and British intellectual prowess, using the examples of Wilson and Maclure, both of whom were born in Scotland.
67. Quoted in Richard D. Wetzel, "Harmonist Music between 1827 and 1832: A Reappraisal," *Communal Studies* 2 (Autumn 1982):78–79.

Bibliography

Manuscript Collections

Bonaparte, Charles Lucien. Scientific Correspondence (microfilm). American Philosophical Society, Philadelphia, Pennsylvania.

Darlington, William. William Darlington Papers, New-York Historical Society, New York, New York.

Engelmann, George. George Engelmann Papers. Missouri Botanical Gardens, St. Louis, Missouri.

Godman, John. Isaac Hays Correspondence. American Philosophical Society, Philadelphia, Pennsylvania.

Gray, Asa. Historic Letter File, Gray Herbarium, Harvard University, Cambridge, Massachusetts.

Harlan, Richard. Miscellaneous Manuscripts. New-York Historical Society, New York, New York.

Harris, Thaddeus. Thaddeus W. Harris Papers. Houghton Library, Harvard University, Cambridge, Massachusetts.

Hicks, Edward. Letters. Galerie St. Etienne, New York, New York.

Hudson, William L. William L. Hudson Journal. American Museum of Natural History, New York, New York.

Maclure, Alexander. Scientific Correspondence (microfilm). American Philosophical Society, Philadelphia, Pennsylvania.

Nuttall, Thomas. Historic Letter File. Gray Herbarium, Harvard University, Cambridge, Massachusetts.

Ord, George. Scientific Correspondence (microfilm). American Philosophical Society, Philadelphia, Pennsylvania.

Owen Family. Owen Family Papers. Manuscript Division, New York Public Library, New York, New York.

Peale, Charles Willson. Charles Willson Peale Papers. American Philosophical Society, Philadelphia.

Museum specimens. Museum of Comparative Zoology and Peabody Museum of Archaeology, Harvard University, Cambridge, Massachusetts.

Peale, Titian Ramsay. Titian R. Peale Papers. American Museum of Natural History, New York.

Titian R. Peale Collection, American Philosophical Society, Philadelphia.

Miscellaneous Correspondence. Museum of Comparative Zoology, Harvard University, Cambridge, Massachusetts.

Titian Ramsay Peale drawings. Philadelphia Academy of Natural Sciences, Philadelphia, Pennsylvania.

Record Unit 139, Smithsonian Institution Archives, Washington, D.C.

Peck, William Dandridge. Manuscript drawings. Houghton Library, Harvard University, Cambridge, Massachusetts.

Rafinesque, C. S. Manuscript drawings. Transylvania University, Lexington, Kentucky.

T. J. Fitzpatrick Collection. Spencer Library, University of Kansas, Lawrence, Kansas.

Say, Lucy. Scientific Correspondence (microfilm). American Philosophical Society, Philadelphia, Pennsylvania.

Thaddeus W. Harris Papers, Houghton Library, Harvard University, Cambridge, Massachusetts.

Miscellaneous Correspondence, Museum of Comparative Zoology, Harvard University, Cambridge, Massachusetts.

Say, Thomas. Thaddeus W. Harris Papers. Houghton Library, Harvard University, Cambridge, Massachusetts.

Manuscript notebook. Museum of Comparative Zoology, Harvard University, Cambridge, Massachusetts.

Melsheimer Correspondence. Philadelphia Academy of Natural Sciences, Philadelphia, Pennsylvania.

Torrey, John. John Torrey Papers. New York Botanical Garden, Bronx, New York.

Wilson, Alexander. Alexander Wilson Papers. Houghton Library, Harvard University, Cambridge, Massachusetts.

Memorabilia. Museum of Comparative Zoology, Harvard University, Cambridge, Massachusetts.

Published Works

Aarsleff, Hans. "An Outline of Language-Origins Theory since the Renaissance." *Annals of the New York Academy of Science* 280 (1976):4–17.

Adams, Alexander B. *John James Audubon.* New York: G. P. Putnam's Sons, 1966.

Adams, Frank D. *The Birth and Development of the Geological Sciences.* New York: Dover, 1954.

Agassiz, Louis. "Synopsis of the Ichthyological Fauna of the Pacific Slope of America." *American Journal of Science,* n.s., 19 (1855):71–99.

Allen, Elsa G. "The History of American Ornithology before Audubon." *Transactions of the American Philosophical Society,* n.s., 41 (1931):387–591.

"American Ornithology." *American Medical and Philosophical Register* 4 (1814):374–579.

"*The American Ornithology* of Alexander Wilson and Charles Lucien Bonaparte." *Edinburgh Evening Post,* November 2, 1832.

"*Animal Kingdom.*" *Monthly American Journal of Geology* 1 (April 1832):447–56.

Appleman, Philip, ed. *Darwin.* New York: W. W. Norton, 1970.

[Arkus, Leon A.] *Hicks, Kane, Pippin.* Pittsburgh: Museum of Art, Carnegie Institute, 1966.

Arndt, Karl. *George Rapp's Harmony Society.* Philadelphia: University of Pennsylvania Press, 1965.

Auden, W. H., and Norman H. Pearson, eds. *Romantic Poets: Blake to Poe.* New York: Viking Press, 1968.

Ayres, James E. "Edward Hicks and His Sources." *Antiques* 109 (February 1976):366–68.

Bachman, John. "An Investigation of the Cases of Hybridity in Animals on Record." *Charleston Medical Journal and Review* (March 1850):168–97.

———. *A Notice of the Types of Mankind.* Charleston: Williams Gettsinger, 1854.

Barham, William. *Descriptions of Niagara, Selected from Various Travellers with Original Additions.* Gravesend: Published by the Compiler, [before 1850].

Barton, Benjamin Smith. *Elements of Botany.* 2 vols. Philadelphia: Printed for the author, 1803.

———. *Fragments of the Natural History of Pennsylvania.* Philadelphia: Printed for the author by Way & Graff, 1799.

———. "An Inquiry into the Question, Whether the Apis Mellifica or True Honey-Bee Is a Native of America." *Transactions of the American Philosophical Society* 3 (1793)241–61.

———. "A Memoir Concerning the Fascinating Faculty Which Has Been Ascribed to the Rattlesnake and Other American Serpents." *Transactions of the American Philosophical Society* 4 (1799):74–113.

Barton, William "Observations on the Probabilities of the Duration of Human Life and the Progress of Populations, in the United States of America." *Transactions of the American Philosophical Society* 3 (1793):25–61.

Bartram, William. "Anecdotes of an American Crow." *Philadelphia Medical and Physical Journal* 1, pt. 1 (1804):89–95.

———. *Botanical and Zoological Drawings.* Edited with an introduction by Joseph Ewan. Philadelphia: American Philosophical Society, 1968.

———. *Travels through North and South Carolina, Georgia, East and West Florida, the Cherokee Country, the Extensive Territories of the Musogulges, or Creek Confederacy, and the Country of the Choctaws: Containing an Account of the Soil and Natural Productions of These Regions, Together with Observations on the Manners.* Philadelphia: James & Johnson, 1791.

Bates, Ralph S. *Scientific Societies in the United States.* Cambridge, Mass.: MIT Press, 1965.

Beidelman, Richard G. "Some Biographical Sidelights on Thomas Nuttall, 1786–1859." *Proceedings of the American Philosophical Society* 104 (1960):86–100.

Beltrami, J. C. *A Pilgrimage in America.* 1828. Reprint. Chicago: Quadrangle Books, n.d.

Berkeley, Edmund, and Dorothy Smith Berkeley. *The Life and Travels of John Bartram.* Tallahassee: University Presses of Florida, 1982.

Bernhard, Karl, Duke of Saxe-Weimar Eisenach. *Travels through North America during the Years 1825 and 1826.* 2 vols. Philadelphia: Carey, Lea and Carey, 1828.

————. *Reise se hoheit des Herzogs Bernhard zu Sachsen-Weimar-Eisenach durch Nord-Amerika indem Jahren 1825 und 1826.* Weimar: bei Wilhelm Hoffmann, 1828.

Bestor, Arthur E. *Backwoods Utopias: The Sectarian and Owenite Phases of Communitarian Socialism in America, 1663–1829.* Philadelphia: University of Pennsylvania Press, 1950.

————. "Education and Reform at New Harmony: Correspondence of William Maclure and Marie DuClos Frétageot, 1820–1833," *Indiana Historical Society Publications* 15 (1948):285–417.

Betts, Edwin M. "The Correspondence between Constantine Samuel Rafinesque and Thomas Jefferson." *Proceedings of the American Philosophical Society* 87 (1943):368–80.

Binney, W. G., ed. *The Complete Writings of Thomas Say on the Conchology of the United States.* 2 vols. New York: H. Baillière, 1858.

Blane, W. N. *Travels through the United States and Canada.* London: Baldwin, 1828.

Boehm, D., and E. Schwartz. "Jefferson and the Theory of Degeneration." *American Quarterly* 9 (1957):454–59.

Boewe, Charles. "Editing Rafinesque Holographs: The Case of the Short Letters." *Filson Club Historical Quarterly* 54 (January 1980):37–49.

Bonaparte, Charles Lucien. *American Ornithology; or, The Natural History of Birds Inhabiting the United States Not Given by Wilson.* 4 vols. Philadelphia: S. A. Mitchell; Carey, Lea, & Carey, 1825–34.

————. "Betrachtungen den Species." *Journal für Ornithologie* 4 (July 1856):257–59.

————. "Further Additions to the Ornithology of the United States and Observations on the Nomenclature of Certain Species." *Annals of the Lyceum of Natural History* 2 (1826–28):154–61.

————. "The Genera of North American Birds and a Synopsis of the Species Found within the Territory of the United States." *Annals of the Lyceum of Natural History* 2 (1826–28):7–128, 293–451.

————. "Observations on the Nomenclature of Wilson's Ornithology." *Journal of the Academy of Natural Science* 3, pt. 2 (1824):349–71; 4, pt. 1 (1824):25–67, 163–200; 4, pt. 2 (1824):251–78; 5, pt. 1 (1825–27):57–106.

Boorstin, Daniel J. *The Lost World of Thomas Jefferson.* New York: Henry Holt, 1948.

Bowler, Peter J. "Bonnet and Buffon: Theories of Generation and the Problem of Species." *Journal of the History of Biology* 6 (1973):259–81.

Bozeman, Theodore Dwight. *Protestants in an Age of Science: The Baconian Ideal and Antebellum American Religious Thought.* Chapel Hill: University of North Carolina Press, 1977.

Brand, Sandra, and Elisa Cullman. *Small Folk: A Celebration of Childhood in America.* New York: E. P. Dutton with the Museum of Folk Art, 1980.

Brown, Roland W. "Jefferson's Contributions to Paleontology." *Journal of the Washington Academy of Science* 83 (1943):257–59.

Browne, Charles A. "Some Relations of the New Harmony Movement to the History of Science in America." *Scientific Monthly* 42 (1936):483–97.

———. "Thomas Jefferson and the Scientific Trends of His Time." *Chronica Botanica* 8 (1944):361–423.

Buckland, William. *"The Birds of America."* *Quarterly Review* 47 (March–July 1832):340.

Buckman, Thomas R., ed. *Bibliography and Natural History.* Lawrence: University of Kansas Libraries, 1966.

Buffon, Georges Louis LeClerc, comte de. *Histoire naturelle, générale et particulière.* 15 vols. Paris: L'imprimerie royale, 1749–67.

Burckhardt, Richard. *The Spirit of System: Lamarck and Evolutionary Biology.* Cambridge, Mass.: Harvard University Press, 1977.

Burham, John C., ed. *Science in America.* New York: Holt, Rinehart and Winston, 1971.

Burns, Frank L. "Miss Lawson's Recollections of Ornithologists." *Auk* 34 (July 1917):275–82.

Cain, Arthur J. "Logic and Memory in Linnaeus' System of Taxonomy." *Proceedings of the Linnaean Society of London* 169 (1956–57):144–63.

Call, Richard E. *The Life and Writings of Rafinesque.* Louisville: John P. Morton, 1895.

Canney, Margaret B. C. *Robert Owen, 1771–1858, Catalogue of an Exhibition of Printed Books, Held in the Library of the University of London, October–December 1958.* London, 1959.

Cassedy, James H. "The Microscope in American Medical Science." *Isis* 67 (1976):76–97.

"Catalogue of the Library of the Academy of Natural Sciences." *Journal of the Academy of Natural Sciences* 1 (1817–18):491–95.

Catlin, George. *Letters and Notes on the Manners, Customs and Conditions of the North American Indians.* 4th ed. 2 vols. London: Published for the author by David Bogue, 1844.

Chamberlain, Montague, ed. *A Popular Handbook of Ornithology of Eastern North America by Thomas Nuttall.* 2d rev. ed. 2 vols. Boston: Little, Brown, 1896.

[Chambers, Robert.] *Vestiges of the Natural History of Creation.* 2 vols. London: John Churchill, 1844.

Chappelsmith, John. "Account of a Tornado Near New Harmony, Indiana, April 30, 1852." *Smithsonian Contributions to Knowledge* 7 (1855):1–68.

Chinard, Gilbert. "The American Philosophical Society and the Early History of Forestry in America." *Proceedings of the American Philosophical Society* 89 (1945):444–87.

———. "The American Sketchbooks of Charles-Alexandre LeSueur." *Proceedings of the American Philosophical Society* 93 (1949):114–18.

———. "Eighteenth Century Theories of America as a Human Habitat." *Proceedings of the American Philosophical Society* 91 (1947):27–57.

Clinton, DeWitt. *An Introductory Discourse Delivered before the Literary and Philosophical Society of New-York on the Fourth of May, 1814.* New York: David Longworth, 1815.

Coates, Benjamin H. *Biographical Sketch of the Late Thomas Say, Esq.* Philadelphia: Academy of Natural Sciences, 1835.

Coleman, William. *Biology in the Nineteenth Century: Problems of Form, Function, and Transformation*. New York: John Wiley & Sons, 1971.

"Comparative View of the Linnaean and Natural Systems of Botany." *Monthly American Journal of Geology* 1 (March 1832):416–22.

Conran, John, ed. *The American Landscape: A Critical Anthology of Prose and Poetry*. New York: Oxford University Press, 1974.

Coonan, Lester P., and Charlotte M. Porter. "Thomas Jefferson and American Biology." *BioScience* 26 (December 1976):745–50.

Cooper, William. "Notices of Big-Bone Lick." *Monthly American Journal of Geology* 1 (1831–32):1–30.

Cooter, R. J. "Phrenology: The Provocation of Progress." *History of Science* 14 (1976):211–34.

Corgan, James X., ed. *The Geological Sciences in the Antebellum South*. University, Ala.: University of Alabama Press, 1982.

Cowdrey, Mary Bartlett. *American Academy of Fine Arts and American Art-Union Exhibition Record, 1816–1852*. New York: New-York Historical Society, 1953.

Cruickshank, Helen G., ed. *John and William Bartram's America*. New York: Devin-Adair, 1957.

Cutright, Paul Russell. *Lewis and Clark: Pioneering Naturalists*. Urbana: University of Illinois Press, 1969.

Dana, E. S., ed. *A Century of Science in America with Special Reference to the American Journal of Science*. New Haven: Yale University Press, 1918.

Daniels, George H. *American Science in the Age of Jackson*. New York: Columbia University Press, 1968.

———. "The Process of Professionalization in American Science, 1820–1860." *Isis* 58 (1967):151–66.

———. "The Pure Science Ideal and Democratic Culture." *Science* 156 (1967):1699–1705.

———, ed. *Darwinism Comes to America*. Waltham: Blaisdell, 1968.

Daniels, George, ed. *Nineteenth Century American Science: A Reappraisal*. Evanston: Northwestern University Press, 1972.

Darlington, William. *Memorials of John Bartram and Humphrey Marshall*. 1849. Facsimile ed., New York: Hafner, 1967.

———. ed. *Reliquiae Baldwinianae*. Philadelphia: Kimber and Sharpless, 1843.

Davidson, Abraham A. *The Eccentrics and Other American Visionary Painters*. New York: E. P. Dutton, 1978.

Davis, John. "The Natural Bridge." *Port Folio* 7, no. 1 (January 1809): 82–84.

Debus, Allen G. *Man and Nature in the Renaissance*. Cambridge: Cambridge University Press, 1978.

DeKay, James E. *Anniversary Address on the Progress of the Natural Sciences in the United States: Delivered before the Lyceum of Natural History of New-York*. New York: G. & C. Carvell, 1826.

Delaporte, François. *Nature's Second Kingdom: Explorations of Vegetality in the Eighteenth Century*. Translated by Arthur Goldhammer. Cambridge, Mass.: MIT Press, 1982.

Dewey, Chester. "Reports on the Herbaceous Plants and Quadrupeds of Massachusetts." *American Journal of Science* 41 (1841):378–81.

Dunlap, William. *The History of the Rise and Progress of the Arts of Design in the United States.* 2 vols. 1834. Reprint. New York: Dover, 1969.

Dupree, A. Hunter. *Asa Gray.* Cambridge, Mass.: Belknap Press of Harvard University Press, 1959.

_____. *Science in the Federal Government.* Cambridge, Mass.: Belknap Press of Harvard University Press, 1957.

_____. "Thomas Nuttall's Controversy with Asa Gray." *Rhodora* 54 (1952):293–303.

Durand, Elias. "Memoir of the Late Thomas Nuttall." *Proceedings of the American Philosophical Society,* n.s., 7 (1861):297–315.

Eaton, Amos. *A Manual of Botany for North America.* 6th ed. Albany: Oliver Steele, 1833.

Echeverria, Durand. *Mirage in the West.* New York: Octagon Books, 1966.

Egerton, Frank N. "Humboldt, Darwin and Population." *Journal of the History of Biology* 3 (Fall 1970):225–59.

Ellicott, Andrew. *Geological Observations in the Mississippi Valley and Florida, 1796–1800.* Philadelphia: Bartram & Budd, 1803.

Elliott, Clark A. *Biographical Dictionary of American Science.* Westport, Conn.: Greenwood Press, 1979.

Emerson, W. Otto. "A Manuscript of Charles Lucien Bonaparte." *Condor* 7 (January–February 1905):44–47.

"The Exhibition." *Port Folio* 8 (August 1812):149–50.

Eyles, Joan M. "G. W. Featherstonhaugh (1780–1866), F.R.S., F.G.S., Geologist and Traveller." *Journal of the Society for the Bibliography of Natural History* 8 (1978):381–95.

Fagin, Nathan B. *William Bartram: Interpreter of the American Landscape.* Baltimore: Johns Hopkins University Press, 1933.

Fairchild, Hermann Neal. *History of the New York Academy of Sciences.* New York: Published by the author, 1887.

Farber, Paul L. "Buffon and the Concept of Species." *Journal of the History of Biology* 5 (1972):259–84.

_____. "The Development of Taxidermy and the History of Ornithology." *Isis* 68 (1977):552–53.

Faxon, Walter. "Relics of Peale's Museum." *Bulletin of the Museum of Comparative Zoology* 59 (July 1915):119–33.

Featherstonhaugh, George. "Eaton's Geology." *Monthly American Journal of Geology* 1 (August 1831):82–90.

_____. *Monthly American Journal of Geology.* Introduction by George W. White. New York: Hafner, 1969.

Fellows, Otis E., and Stephen F. Milliken. *Buffon.* New York: Twayne, 1972.

Fitzpatrick, Thomas J. *Rafinesque: A Sketch of His Life with Bibliography.* Des Moines: Historical Department of Iowa, 1911.

Flexner, James. *The Light of Distant Skies.* New York: Dover, 1969.

Ford, Alice. *Edward Hicks: Painter of the Peaceable Kingdom.* Philadelphia: University of Pennsylvania, 1952.

_____. *John James Audubon.* Norman: University of Oklahoma Press, 1964.

_____. "The Publication of Edward Hicks' Memoirs." *Bulletin of the Friends Historical Association* 50 (Spring 1961):5.

Fox, W. J. "Letters from Thomas Say to John F. Melsheimer, 1816–1825." *Entomological News and Proceedings of the Entomological Section of the Academy of Natural Sciences* 12 (1901):110–12, 138–41, 173–77, 203–5, 233–36, 281–83, 314–16.

Frangsmayr, Tore, ed. *Linnaeus: The Man and His Work.* Berkeley and Los Angeles: University of California Press, 1983.

Friedman, Gerald M. " 'Gems' from Rensselaer." *Earth Sciences History* 2 (1983):97–102.

Fries, Waldemar H. *The Double Elephant Folio.* Chicago: American Library Association, 1973.

Fuller, Harlan M., and LeRoy R. Hafen, eds. *The Journal of Captain R. Bell.* Glendale, Calif.: Arthur H. Clark, 1957.

Fulling, Edmund H. "Thomas Jefferson: His Interest in Plant Life as Revealed in His Writings, II." *Bulletin of the Torrey Botanical Club* 72 (May 1945): 248–70.

Gager, C. Stuart. "Botanic Gardens in Science and Education." *Science* 85 (1937):393–99.

Gay, Peter. *The Party of Humanity.* New York: W. W. Norton, 1959.

Gerstner, Patsy. "Vertebrate Paleontology, an Early Nineteenth-Century Transatlantic Science." *Journal of the History of Biology* 3 (1970):137–48.

Gibson, R. J. Harvey. *Outlines of the History of Botany.* London: A. & C. Black, 1919.

Gilmer, Francis W. "On the Geological Formation of the Natural Bridge of Virginia." *Transactions of the American Philosophical Society,* n.s., 1 (1818):187–92.

Godman, John. *American Natural History.* 3 vols. Philadelphia: H. C. Carey and I. Lea, 1826–28.

Goetzmann, William H. *Exploration and Empire: The Explorer and the Scientist in the Winning of the American West.* New York: Alfred A. Knopf, 1971.

Goode, G. B. "The Beginnings of Natural History in America." *Annual Report of the Smithsonian Institution, National Museum.* Pt. 2. Washington, D.C.: U.S. Government Printing Office, 1901.

Graustein, Jeannette E. "Audubon and Nuttall." *Scientific Monthly* 74 (1952):84–90.

―――. *Thomas Nuttall, Naturalist.* Cambridge, Mass.: Harvard University Press, 1967.

[Gray, Asa]. "The Longevity of Trees." *North American Review* 59 (1844):190–238.

―――. "A Natural System of Botany." *American Journal of Science* 32 (July 1837):292–303.

―――. "Notice of the Botanical Writings of the Late C. S. Rafinesque." *American Journal of Science* 40 (January–March 1841):221–41.

Greene, John C. "American Science Comes of Age, 1780–1820." *Journal of American History* 60 (1968):22–41.

―――. "The Development of Mineralogy in Philadelphia." *Proceedings of the American Philosophical Society* 113 (1969):283–95.

Greenwood, F. W. P. "An Address Delivered before the Boston Society of Natural History." *Boston Journal of Natural History* 1 (May 1834):7–14.

Grosart, Alexander B. *The Poems and Literary Prose of Alexander Wilson.* 2 vols. Paisley, Scotland: Alex. Gardner, 1876.

Gross, Samuel D. *Autobiography of Samuel D. Gross, M.D.* Philadelphia: George Barrie, 1887.

Gummere, Richard M. *The American Colonial Mind and the Classical Tradition.* Cambridge, Mass.: Harvard University Press, 1963.

Guralnick, Stanley M. *Science and the Ante-Bellum American College.* Philadelphia: American Philosophical Society, 1975.

_____. "Sources of Misconception on the Roles of Science in the Nineteenth-Century American College." *Isis* 65 (1975):352–66.

Guthrie, William, ed. *A New Geographical, Historical and Commercial Grammar.* Philadelphia: Johnson & Walker, 1815.

Haldeman, S. S. "Notice of the Zoological Writings of the Late C. S. Rafinesque." *American Journal of Science* 42 (1842):280–91.

_____. *On the Impropriety of Using Vulgar Names in Zoology.* Philadelphia: Carey and Hart, Judiah Dobson, and John Pennington; New York: Wiley and Putnam, 1843.

Hall, Courtney Robert. *A Scientist in the Early Republic.* New York: Columbia University Press, 1934.

Hall, Thomas, ed. *A Source Book in Animal Biology.* Cambridge, Mass.: Harvard University Press, 1970.

Hanley, Wayne. *Natural History in America.* New York: Quadrangle, New York Times Book Co., 1977.

Harlan, Richard. "Description of an Hermaphrodite Orang Outang, Lately Living in Philadelphia." *Journal of the Academy of Natural Sciences* 5 (1825–27):229–36.

_____. "Description of *Vespertilio Auduboni*, a New Species of Bat." *Monthly American Journal of Geology* 1 (November 1831):217–21.

_____. *Fauna Americana, Being a Description of the Mammiferous Animals Inhabiting North America.* Philadelphia: Anthony Finley, 1825.

_____. *Medical and Physical Researches.* Philadelphia: Lydia R. Bailey. 1835.

_____. *Refutation of Certain Misrepresentations Issued against the Author of "Fauna Americana."* . . . Philadelphia: William Stavely, 1826.

Harper, Francis, ed. *The Travels of William Bartram: Naturalist's Edition.* New Haven: Yale University Press, 1958.

Haskell, Daniel C. *The United States Exploring Expedition, 1838–1842, and Its Publications, 1844–1874.* New York: New York Public Library, 1942.

Hatch, Melville. "Coleoptera." *A Century of Progress in the Natural Sciences, 1853–1953.* San Francisco: California Academy of Sciences, 1955.

Henrey, Blanche. *British Botanical and Horticultural Literature before 1800.* 3 vols. London: Oxford University Press, 1975.

Herrick, Francis H. *Audubon the Naturalist.* 2 vols. New York: D. D. Appleton, 1917.

Hibernicus [DeWitt Clinton]. *Letters on the Natural History and Internal Resources of the State of New York.* New York: Sold by E. Bliss & E. White, 1822.

Hicks, Edward. *Memoirs of the Life and Religious Labors of Edward Hicks, Late of Newtown, Bucks County, Pennsylvania, Written by Himself.* Philadelphia: Merrihew & Thompson, 1851.

Himmelfarb, Gertrude. *Darwin and the Darwinian Revolution.* New York: W. W. Norton, 1962.

Hindle, Brooke. *The Pursuit of Science in Revolutionary America, 1735–1789.* 1956. Reprint. New York: W. W. Norton, 1974.

[Hitchcock, Edward.] "The New Theory of the Earth." *North American Review* 28 (1829):265–94.

Holthuis, L. B., and M. Boeseman. "Notes of C. S. Rafinesque-Schmaltz's (1810) Caratteri di alcuni nuovi generi e nuove specie di animali e pianti della Sicilia." *Journal of the Society for the Bibliography of Natural History* 8 (1977):231–34.

Horsman, Reginald. "Scientific Racism and the American Indian in the Mid-Nineteenth Century," *American Quarterly* 27 (1975):152–65.

Horstadius, Sven. "Linnaeus, Animals and Man." *Biological Journal of the Linnaean Society* 6 (December 1974):269–75.

Humphreys, William J. "A Review of Papers in Meteorology and Climatology Published by the American Philosophical Society prior to the Twentieth Century." *Proceedings of the American Philosophical Society* 86 (1942):29–33.

Huth, Hans. *Nature and the American.* Lincoln, Neb.: Bison Book, 1972.

Huxley, Leonard, ed. *Life and Letters of Joseph Dalton Hooker.* 2 vols. London: J. Murray, 1918.

Jackson, Donald, ed. *Letters of the Lewis and Clark Expedition with Related Documents.* Urbana: University of Illinois Press, 1962.

Jaffe, Bernard. *Men of Science.* New York: Simon & Schuster, 1958.

James, Edwin. *Account of an Expedition from Pittsburgh to the Rocky Mountains, Performed in the Years 1819, 1820.* 2 vols. and atlas. Philadelphia: H. C. Carey and I. Lea, 1823.

———. *Account of an Expedition from Pittsburgh to the Rocky Mountains Performed in the Years 1819, 1820.* 3 vols. London: Hurst, Rees, Orme, & Brown, 1823.

Jameson, Robert, ed. *Constable's Miscellany of Original and Selected Publications in Various Departments of Literature, Science, and the Arts.* Vols. 71–74. Edinburgh: Constable, 1831; London: Hurst, Chance and Co., 1831.

Jefferson, Thomas. "A Memoir on the Discovery of Certain Bones of a Quadruped of the Clawed Kind in the Western Parts of Virginia." *Proceedings of the American Philosophical Society* 4 (1799):246–60.

———. *Notes on the State of Virginia.* Philadelphia: R. T. Rawle, 1801.

Jennings, Francis. *The Invasion of America: Indians, Colonialism, and the Cant of Conquest.* Chapel Hill: University of North Carolina Press, 1975.

Johnson, Markes E. "Geology and Early American Reforms in Education: The Rensselaer and New Harmony Schools." Ph.D. dissertation, Williams College, 1980.

Jones, Howard Mumford. *America and French Culture, 1750–1848.* Chapel Hill: University of North Carolina Press, 1927.

Jordan, David Starr. "Rafinesque." *Popular Science Monthly* 29 (June 1886):212–21.

Kallir, Jane. *The Folk Art Tradition: Naive Painting in Europe and the United States.* New York: Viking Press, 1981.

Kastner, Joseph. *A Species of Eternity.* New York: Alfred A. Knopf, 1977.

Keating, William. *Narrative of an Expedition to the Source of St. Peter's River, Lake Winnepeck, Lake of the Woods, &c, &c.* Philadelphia: H. C. Carey and I. Lea, 1824.

Lapsansky, Emma Jones. "'Since They Got Those Separate Churches': Afro-Americans and Racism in Jacksonian Philadelphia." *American Quarterly* 32 (Spring 1980):54–78.

Latrobe, Benjamin H. "A Drawing and Description of the Clupea Tyrranus and Oniscus Praegustator." *Transactions of the American Philosophical Society* 5 (1802):77–81.

Lawrence, William. *Lectures on Physiology, Zoology and the Natural History of Man.* Salem: Foote & Brown, 1828.

LeConte, John L., ed. *The Complete Writings of Thomas Say on the Entomology of North America.* 2 vols. New York: Baillière Brothers, 1859.

Leidy, Joseph. "A Memoir on the Extinct Sloth Tribe." *Smithsonian Contributions to Knowledge* 7 (1855):1–68.

Leighton, Ann. *American Gardens in the Eighteenth Century.* Boston: Houghton, Mifflin, 1976.

Leopold, Richard W. *Robert Dale Owen.* Cambridge, Mass.: Harvard University Press, 1940.

Lincoln, Almira. *Familiar Lectures on Botany.* New York: F. J. Huntington and Mason and Law, 1852.

Lindley, John. *An Introduction to the Natural System of Botany or a Systematic View of the Organization, Natural Affinities and Geographic Distributions of the Whole Vegetable Kingdom with a Catalogue of North American Genera of Plants* by John Torrey. London: 1830; New York: G. and C. and H. Carvill, 1831.

———. *The Vegetable Kingdom.* 3d ed. London: Bradbury & Evans, 1853.

Lipman, Jean, and Tom Armstrong, eds. *American Folk Painters of Three Centuries.* New York: Hudson Hills Press with the Whitney Museum of American Art, 1980.

"Literary Intelligence." *Port Folio* 8 (November 1819):434.

Lockridge, Ross R. *The Old Fauntleroy Home.* Published for the New Harmony Memorial Commission by Mrs. Edmund Burke Ball, 1939.

Lockwood, George B. *The New Harmony Communities.* Marion, Ind.: Chronicle Co., 1902.

Lockwood, George B., and Charles Prosser. *The New Harmony Experiment.* New York: Appleton, 1905.

Loring, Jane, ed. *Letters of Asa Gray.* 2 vols. Boston: Houghton, Mifflin, 1894.

Lyell, Charles. *Principles of Geology.* Philadelphia: James Kay, Jr., and Brother, 1837.

"Lyell's Geology: First American Edition from the First and Last London Edition." *American Journal of Science* 32 (July 1837):182–83.

Lyon, John and Philip R. Sloan. *From Natural History to a History of Nature.* Notre Dame: University of Notre Dame Press, 1981.

McAllister, Ethel M. *Amos Eaton, Scientist and Educator, 1776–1842.* Philadelphia: University of Pennsylvania Press, 1941.

McDermott, Francis, ed. *Audubon in the West.* Norman: University of Oklahoma Press, 1965.

Maclure, William. "Observations on the Geology of the United States, Explanatory of a Geologic Map." *Transactions of the American Philosophical Society* 6 (1809):411–28.

———. *Observations on the Geology of the United States of America.* Philadelphia: Abraham Small, 1817.

———. *Opinions on Various Subjects.* 3 vols. New Harmony: School Press, 1831–38.

Mallis, Arnold. *American Entomologists.* New Brunswick: Rutgers University Press, 1971.

Mann, Maybelle. "American Landscape Prints." *Arts & Antiques.* 4 (May–June 1981):90–97.

Martin, E. T. *Thomas Jefferson: Scientist.* New York: Henry Schuman, 1952.

Martin, Paul S. "The Discovery of America." *Science* 179 (March 1973):969–74.

Maximilian, Alexander Philip, Prince of Wied-Neuwied. *Travels in the Interior of North America, 1832–1834.* 2 vols. and atlas. Translated by H. Evans Lloyd. London: Ackermann, 1843.

Meigs, Charles D. *A Memoir of Samuel George Morton.* Philadelphia: T. K. & P. G. Collins, 1851.

Merrill, Elmer D. "A Generally Overlooked Rafinesque Paper." *Proceedings of the American Philosophical Society* 86 (1942):72–90.

———. "A Life of Travels by C. S. Rafinesque." *Chronica Botanica* 8 (1944):292–360.

———. "Rafinesque's Publications from the Standpoint of World Botany." *Proceedings of the American Philosophical Society* 87 (1943):110–19.

Merrill, George P. "Contributions to a History of American State Geological and Natural History Surveys." *Bulletin of the United States National Museum* 109 (1920):149–58.

———. *The First One Hundred Years of American Geology.* New Haven: Yale University Press, 1924.

———. "The *Monthly American Journal of Geology and Natural Sciences.*" *American Geology* 30 (July–December 1902):62–64.

Michaux, François André. *The North American Sylva.* Translated by Augustus L. Hillhouse. 3 vols. Paris: C. D. Hautel for T. Dobson and G. Conrad of Philadelphia, 1818, 1819.

———. *The North American Sylva.* Edited by J. J. Smith, 2 vols. Philadelphia: Rice, Rutter & Co., 1865.

Milcher, Marilyn S. "Round Panel Furniture of Virginia's Eastern Shore, 1730–1830." *Art & Antiques* 5 (November–December 1982):85.

[Mitchill, Samuel Latham.] *"American Ornithology, or the Natural History of the Birds of the United States." Medical Repository* 14 (1811):48.

———. *"American Ornithology, or the Natural History of the Birds of the United States." Medical Repository* 17 (1814):250.

———. *Discourse on Thomas Jefferson More Especially as a Promoter of Natural and Physical Sciences.* New York: G. C. Carvill, 1826.

———. *Essay on the Theory of the Earth by M. Cuvier with Mineralogical Notes and an Account of Cuvier's Geological Discoveries by Professor Jameson to Which is Now Added Observations on the Geology of North America.* New York: Kirk and Mercein, 1818.

_____. *Observations on the Geology of North America.* New York: Kirk and Mercein, 1818.

_____. *The Picture of New York, or, The Traveller's Guide through the Commercial Metropolis of the United States.* New York: Riley, 1807.

_____. *Report in Part on the Fishes of New York.* Edited by Theodore Gill. Washington, D.C.: Printed for the editor, 1898.

Möllhausen, Baldwin. *Diary of a Journey from the Mississippi to the Coast of the Pacific with a United States Government Expedition.* Translated by Mrs. Percy Sinnett. 2 vols. London: Longman, Brown, Green, Longmans & Roberts, 1858.

Monroe, W. S. *History of the Pestalozzian Movement in the United States.* Syracuse: Bardeen, 1907.

Montagu, M. F. Ashley. *Edward Tyson, M.D., F.R.S. (1650–1708) and the Rise of Comparative Anatomy in England.* Philadelphia: American Philosophical Society, 1943.

_____. ,ed. *Studies in the History of Science and Learning.* 1944. Reprint. New York: Krause, 1969.

Moore, J. Percy. "William Maclure—Scientist and Humanitarian." *Proceedings of the American Philosophical Society* 91 (1947):234–49.

Morris, Julia Lewis. *From Seed to Flower.* Philadelphia: Pennsylvania Horticultural Society, 1976.

Morton, Samuel George. *Brief Remarks on the Diversities of the Human Race and Some Kindred Subjects.* Philadelphia: Messchew & Thompson, 1842.

_____. *Catalogue of Skulls of Man and the Inferior Animals in the Collection of Samuel G. Morton.* Philadelphia: Turner and Fisher, 1840.

_____. *Crania Americana, or a Comparative View of the Skulls of Various Aboriginal Nations of North and South America.* Philadelphia: J. Dobson, 1839.

_____. "History of the Academy of Natural Sciences of Philadelphia." *American Quarterly Register* 13 (1841):433–38.

_____. *Letter to John Bachman on the Question of Hybridity in Animals.* Charleston: Walker & Johnson, 1850.

_____. *A Memoir of William Maclure.* Philadelphia: T. K. & P. G. Collins, 1841.

_____. *Notice of the Academy of Natural Sciences of Philadelphia.* 2d ed. Philadelphia: Mifflin and Parry, 1831.

Mosimann, James E., and Paul S. Martin. "Simulating Overkill by Paleo Indians." *American Science* 63 (May–June 1975):304–13.

Mumford, Lewis. *Sticks and Stones.* 2d rev. ed. New York: Dover, 1955.

Natural History of Animals, Vegetables and Minerals with the Theory of Earth in General. 6 vols. London: Printed for T. Bell, 1775.

Nelson, Gareth. "From Candolle to Croizat: Comments on the History of Biogeography." *Journal of the History of Biology* 11 (Fall 1978):269–305.

"The New York Academy of Sciences and the American Intellectual Tradition: An Historical Overview." *Transactions of the New York Academy of Sciences,* 2d ser., 37 (1975):3–34.

Nichols, Roger L., ed. *The Missouri Expedition (1818–1820).* Norman: University of Oklahoma Press, 1969.

_____. "Stephen Long and Scientific Explorations on the Plains." *Nebraska History* 52 (Spring 1971):51–64.

Nicholson, William, ed. *American Edition of the British Encyclopedia or Dictionary of Arts and Sciences.* 12 vols. Philadelphia: Mitchell, Ames, and White, 1819.

Nicolson, Marjorie H. *Mountain Gloom and Mountain Glory.* Ithaca: Cornell University Press, 1959.

Nordenskiöld, Erik. *The History of Biology.* New York: Tudor, 1935.

"Notice and Review of the 'Reliquiae Diluviae.' Or Observations on the Organic Remains Contained in Caves, Fissures and Diluvial Gravel, and on Other Geological Phenomena Attesting the Action of an Universal Deluge." *American Journal of Science* 8 (August 1824):150–68.

Nuttall, Thomas. *The Genera of North American Plants and Catalogue of the Species to the Year 1817.* 2 vols. Philadelphia: D. Heartt, 1818.

———. *An Introduction to Systematic and Physiological Botany.* Cambridge: Hilliard and Brown, Booksellers to the University, 1827. 2d ed., 1830.

———. *The North American Sylva of F. Andrew Michaux.* 3 vols. Philadelphia: J. Dobson, 1842–52.

Nye, Ronald. *The Cultural Life of a New Nation.* New York: Harper & Row, 1960.

Oleson, Alexandra, and Sanborn C. Brown, eds. *The Pursuit of Knowledge in the Early American Republic.* Baltimore: Johns Hopkins University Press, 1976.

"On the Causes Which Retard the Advances of Zoological Knowledge." *Monthly American Journal of Geology* 1 (January 1832):302.

Ord, George. "A Memoir of Charles-Alexandre LeSueur." *American Journal of Science,* 2d ser., 8 (1849):189–216.

———. *North American Zoology.* Edited by Samuel N. Rhoads. Haddonfield: Printed for the author, 1894.

Ord, George, and Thomas Say. "A New Genus of Mammalia Proposed, and Description of the Species upon Which It Is Founded." *Journal of Academy of Natural Sciences* 5 (1825):340–50.

"Original Letter from Sir Benjamin West to Charles W. Peale, Esq." *Port Folio* 3 (July 1810):8–13.

Osborn, Henry. "Thomas Jefferson as a Paleontologist." *Science* 82 (1935):533–37.

Owen, Robert. *A New View of Society; or, Essays on the Principle of the Formation of the Human Character, and the Application of the Principle to Practice.* London: Cadell and Davies, Strand, 1813.

Pammell, L. H. "Dr. Edwin James." *Annals of Iowa.* 3d ser., 8 (1907):161–75; 277–95.

Pancoast, Elinor. *The Incorrigible Idealist: Robert Dale Owen in America.* Bloomington: Principia Press, 1940.

Peale, Charles Willson. *Discourse Introductory to a Course of Lectures in the Science of Nature with Original Music.* Philadelphia: Zachariah Poulson, 1800.

———. *Scientific and Descriptive Catalogue of Peale's Museum.* Philadelphia: Samuel H. Smith, 1796.

Peale, Titian R. *Lepidoptera Americana; or, Original Figures of Moths and Butterflies of North America: In Their Various Stages of Existence and the Plants on Which They Feed.* Prospectus. Philadelphia: William P. Gibbons. 1833.

Peattie, Donald C. *Green Laurels.* New York: Literary Guild, 1936.

Peck, William Dandridge. "Four Remarkable Fishes Taken Near the Piscotaqua, N.H." *Memoirs of the American Academy of Arts and Sciences* 2 (1797):46.

Pennell, Francis W. "Benjamin Smith Barton as a Naturalist." *Proceedings of the American Philosophical Society* 86 (1942):108–22.

———. "Travels and Scientific Collections of Thomas Nuttall." *Bartonia*, no. 18 (December 1936), pp. 1–51.

Phillips, Maurice E. "The Academy of Natural Sciences of Philadelphia." *Transactions of the American Philosophical Society*, n.s., 43 (1953):266–71.

Pickering, Charles. *Chronological History of Plants: Man's Record of His Own Existence Illustrated through Their Names, Uses, and Companionship*. Boston: Little, Brown, 1879.

———. "On the Geographical Distribution of Plants." *Transactions of the American Philosophical Society* n.s., 3 (1830):274–84.

———. *Races of Man: And Their Geographical Distribution*. Boston: Charles C. Little & James Brown, 1848.

Poesch, Jessie. *Titian Ramsay Peale and His Journals of the Wilkes Expedition*. Philadelphia: American Philosophical Society, 1961.

Porter, Charlotte M. "The American West Described in Natural History Journals, 1819–1836." *Midwest Review* 2 (1980):25–26.

———. "The Excursive Naturalists: The Development of American Taxonomy at the Philadelphia Academy of Natural Sciences, 1812–1842." Ph.D. dissertation, Harvard University, 1976.

———. "Following Bartram's 'Track': Titian Ramsay Peale's Florida Journey." *Florida Historical Quarterly* 61 (1983):431–34.

———. "The Lifework of Titian Ramsay Peale." *Proceedings of the American Philosophical Society* 129 (1985):300–12.

———. "'Subsilentio': Discouraged Works of American Natural History." *Journal of the Society for the Bibliography of Natural History* 9 (1979):109–19.

Porter, H. H. "Preview of the Ornithological Biography." *Monthly American Journal of Geology* 1 (September 1831):136–39.

Poulson, C. A. *A Monograph of the Fluviatile Bivalve Shells of the River Ohio*. Philadelphia: J. Dobson, 1832.

———. "Rafinesque's Monograph of the Bivalve Shells of the River Ohio." *Monthly American Journal of Geology and Natural Sciences* 1 (February 1832):370–77.

Prest, John. *The Garden of Eden*. New Haven: Yale University Press, 1981.

Pursell, Carroll, Jr., ed. *Technology in America*. Cambridge, Mass.: MIT Press, 1982.

Quinn, Edward. "Thomas Jefferson and the Fossil Record." *Bios* 47 (December 1976):161–64.

Rafinesque, Constantine S. *Autikon Botanikon*. Philadelphia, 1815, 1840.

———. *Continuation of a Monograph of the Bivalve Shells of the River Ohio*. Philadelphia, 1831.

———. *Flora Telluriana*. Philadelphia: Printed for the author by H. Probasco, 1836.

———. "Fragments of a Letter to Mr. Bory St. Vincent at Paris." *Western Minerva* 1 (1821):70–71.

———. *Herbarium Rafinesquiarum*, Prodromus. Philadelphia: S. C. Beck, 1833.

———. *Ichthyologia Ohiensis*. Lexington: Printed for the author by W. G. Hunt, 1820.

————. *A Life of Travels*. Philadelphia: Printed for the author, 1836.

————. *New Flora and Botany of North America*. 1836. Reprint. Cambridge, Mass.: Murray, 1946.

————. *On Botany (1820)*. Edited by Charles Boewe. Frankfort: Whippoorwill Press, 1983.

————. *The Pleasures and Duties of Wealth*. Philadelphia: Eleutherium of Knowledge, 1840.

————. "Principles of the Philosophy on New Genera and New Species of Plants and Animals." *Atlantic Journal* 1 (1832–33):163–64.

————. *The World or Instability: A Poem in Twenty Parts*. Philadelphia: J. Dobson; London: O. Rich, 1836.

Ramsay, David. "An Oration on the Advantages of American Independence." *United States Magazine* 1 (1779):53.

[Ray, I.] "DeCandolle's Botany." *North American Review* 38 (1834):32–62.

Reese, Rena, ed. "List of Books and Pamphlets in the Library of the Workingman's Institute, New Harmony, Indiana," New Harmony, March 1909. Mimeograph, Widener Library, Harvard University.

Reingold, Nathan. "Reflections on 200 Years of Science in the United States." *Nature* 262 (1976):9–13.

————, ed. *Science in America since 1820*. New York: Science History Publications, 1976.

————, ed. *Science in Nineteenth Century America: A Documentary History*. New York: Hill & Wang, 1964.

Reinhold, Meyer. "The Quest for 'Useful Knowledge' in Eighteenth-Century America." *Proceedings of the American Philosophical Society* 119 (April 1975):108–32.

"Remarks of the Publishing Committee." *Journal of the Academy of Natural Sciences* 1, pt. 2 (1818):485–86.

Rhoads, Samuel N. "George Ord." *Cassina*, no. 12 (1908), pp. 1–8.

Richardson, Edgar P., Brooke Hindle, and Lillian B. Miller. *Charles Willson Peale and His World*. New York: Harry N. Abrams, 1982.

Rodgers, Andrew D. *John Torrey: A Story of North American Botany*. New York: Hafner, 1965.

Root, N. J. "The Rare Book and Manuscript Collection of the American Museum of Natural History Library." *Curator* 20 (1977):121–52.

Rush, Benjamin. *Medical Inquiries and Observations upon the Diseases of the Mind*. Philadelphia: Kimber & Richardson, 1812.

————. "Observations Intended to Favor a Supposition That the Black Color (as It Is Called) of the Negroes Is Derived from the Leprosy." *Transactions of the American Philosophical Society* 4 (1799):289–97.

Russett, Cynthia Eagle. *Darwin in America*. San Francisco: W. H. Freeman, 1976.

Sach, Julius von. *History of Botany*. Translated by Henry E. F. Garnsey. Revised by J. B. Balfour. Oxford: Clarendon Press, 1906.

Sage, John H. "An Historic Letter." *Auk* 7 (October 1895):359–62.

Say, Thomas. *American Entomology, or Descriptions of the Insects of North America*. 3 vols. Philadelphia: S. A. Mitchell, 1824, 1825, 1828.

————. *Glossary to Say's Entomology*. Philadelphia: S. A. Mitchell, 1825.

_____. *Glossary to Say's Conchology.* New Harmony, Ind.: Richard Beck and James Bennett, 1832.

Schenck, Jacob, and Richard Owen. *The Rappites: Interesting Notes about New Harmony.* Evansville: Courier Co., 1890.

Schmaltz, C. Rafinesque. "Progress in American Botany." *Medical Repository* 1 (1810):297.

Schwartz, Joel S. "Charles Darwin's Debt to Malthus and Edward Blythe." *Journal of the History of Biology* 7 (Fall 1974):301–18.

Sellers, Charles Coleman. *Charles Willson Peale.* 2 vols. Philadelphia: American Philosophical Society, 1947.

_____. *Mr. Peale's Museum.* New York: W. W. Norton, 1980.

_____. "Portraits and Miniatures by Charles Willson Peale." *Transactions of the American Philosophical Society* 42 (June 1952):3–369.

Shapin, Steven. "The Pioneer Days of Progress." *Times Literary Supplement,* January 7, 1977, p. 3.

Shear, C. L., and Neil E. Stevens, eds. "The Correspondence of Schweinitz and Torrey." *Memoirs of the Torrey Botanical Club* 16 (1915–21):119–300.

Sheets-Pyenson, Susan. "War and Peace in Natural History Publications: The Naturalist's Library, 1833–1842." *Isis* 72 (March 1981):50–72.

Silliman, Benjamin. "Extracts from Letters Addressed to the Editor by William Maclure." *American Journal of Science* 9 (1825):157–64.

_____. "Flora of the Middle and Northern States." *American Journal of Science* 8 (1824):178.

_____. "Memoranda, Extracted from a Letter to the Editor, Dated Alicante (Spain) March 6, 1824 from William Maclure." *American Journal of Science* 8 (1824):187–90.

_____. "Mr. Owen and His Plan of Education." *American Journal of Science* 9 (June 1825):383–84.

Simpson, George Gaylord. "The Beginnings of Vertebrate Paleontology in North America." *Proceedings of the American Philosophical Society* 86 (1942):130–57.

Smallwood, William. "Amos Eaton, Naturalist." *New York History* 18 (1937):167–88.

Smellie, William. *Natural History, General and Particular.* 9 vols. London: W. Strahan and T. Cadell; Edinburgh: W. Creech, 1770; 2d ed., 1785.

Smith, James E., ed. *A Selection of the Correspondence of Linnaeus and Other Naturalists.* 2 vols. London: Longman, Hurst, Rees, Orme, and Brown, 1821.

Sparks, Jared, comp. *The Library of American Biography.* 25 vols. Boston: Hilliard, Gray, 1834–48.

_____. "Wilson's and Bonaparte's Ornithology," *North American Review* 24 (January 1827):110–29.

Spencer, Frank. "Two Unpublished Essays on the Anthropology of North America by Benjamin Smith Barton." *Isis* 68 (December 1977):567–73.

Stanton, William. *The Great United States Exploring Expedition of 1838–1842.* Berkeley and Los Angeles: University of California Press, 1975.

_____. *The Leopard's Spots: Scientific Attitudes toward Race in America, 1815–1859.* Chicago: University of Chicago Press, 1960.

Stearns, Raymond Phineas. *Science in the British Colonies of America.* Urbana: University of Illinois Press, 1970.

Sterki, Victor. "A Few Notes on Say's Early Writings and Species." *Nautilus* 21 (1907):31–34.

Stresemann, Erwin. *Ornithology from Aristotle to the Present.* Translated by H. J. Epstein and Cathleen Epstein. Cambridge, Mass.: Harvard University Press, 1975.

Struik, Dirk. *Yankee Science in the Making.* Boston: Little, Brown, 1948.

Sumner, George. *A Compendium of Physiological and Systematic Botany.* Hartford: Oliver D. Cooke, 1820.

Swainson, William. *A Preliminary Discourse on the Study of Natural History.* London: Longman, Rees, Orme, Brown, Green and Longman, 1834.

Tanner, H. S. *Description of the Canals and Rail Roads of the United States in 1840.* New York: T. R. Tanner, 1844.

Tatham, David. "Edward Hicks, Elias Hicks and John Comly: Perspectives on the Peaceable Kingdom Theme." *American Art Journal* 13 (Spring 1981):47–48.

Taton, René, ed. *The Beginnings of Modern Science from 1450 to 1800.* 2 vols. Translated by A. J. Pomerans. New York: Basic Books, 1966.

Tebeau, Charlton W. *A History of Florida.* Coral Gables: University of Miami Press, 1971.

Thoreson, Timothy H. H., ed. *Toward a Science of Man: Essays in the History of Anthropology.* The Hague: Mouton, 1975.

Thwaites, Reuben Gold, ed. *Early Western Travels.* 32 vols. Cleveland: Arthur H. Clark, 1904–7.

Tocqueville, Alexis de. *Democracy in America.* Translated by George Lawrence. New York: Doubleday, 1969.

Tolles, Frederick B. "The Primitive Painter as Poet." *Bulletin of the Friends Historical Association* 50 (Spring 1969):12–30.

Torrey, John. *Flora of the Middle and Northern Sections of the United States; or, a Systematic Arrangement and Description of All the Plants Hitherto Discovered in the United States North of Virginia.* 2 vols. New York: T. and J. Swords, 1823–24.

———. "Some Account of a Collection of Plants." *Annals of the Lyceum of Natural History* 2 (1826–28):161–254.

Townsend, John K. *Narrative of a Journey across the Rocky Mountains to the Columbia River, and a Visit to the Sandwich Islands, Chili &c.* Philadelphia: Henry Perkins, 1839.

Trautman, Milton B. *The Fishes of Ohio.* Baltimore: Waverly Press, 1957.

Turner, George. "Memoir on the Extraneous Fossils Denominated Mammoth Bones . . ." *Proceedings of the American Philosophical Society* 4 (1799):510–18.

Vail, R. W. G. *The American Sketchbooks of Charles-Alexandre LeSueur.* Worcester: American Antiquarian Society, 1938.

Verner, Coolie. *A Further Checklist on the Separate Editions of Jefferson's Notes on the State of Virginia.* Charlottesville: Bibliographical Society of the University of Virginia, 1950.

———. *Mr. Jefferson Distributes His Notes.* New York: New York Public Library, 1932.

Vigors, Nicholas A. "Observations on the Natural Affinities That Connect the Orders and Families of Birds." *Transactions of the Linnaean Society of London* 14 (1825):395–517.

Vorzimmer, Peter. "Darwin, Malthus and the Theory of Natural Selection." *Journal of the History of Ideas* 30 (1969):527–42.

Walker, Robert H. *The Poet and the Gilded Age: Social Themes in Late Nineteenth Century American Verse.* Philadelphia: University of Pennsylvania Press, 1963.

Warner, Deborah J. "Science Education for Women in Antebellum America." *Isis* 69 (1978):58–67.

Weese, A. O., ed. "The Journal of Titian Ramsay Peale, Pioneer Naturalist." *Missouri Historical Review* 41 (1947):147–63, 266–84.

Weiss, Harry B., and Grace M. Ziegler. "The Communism of Thomas Say." *Journal of the New York Entomological Society* 35 (September 1937):231–39.

———. *Thomas Say, Early American Naturalist.* Baltimore: Charles C. Thomas, 1931.

Welker, Robert H. *Birds and Men: American Birds in Science, Art, Literature, and Conservation, 1800–1900.* Cambridge, Mass.: Harvard University Press, 1955; New York: Atheneum, 1966.

Wells, Kentwood D. "Sir William Lawrence (1783–1867): A Study of Pre-Darwinian Ideas on Heredity and Variation." *Journal of the History of Biology* 4 (1971):319–61.

Wetzel, Richard D. "Harmonist Music between 1827 and 1832: A Reappraisal." *Communal Studies* 2 (Autumn 1982):65–84.

Wheeler, Alwyne, ed. *Contributions to the History of North American Natural History.* London: Society for the Bibliography of Natural History, 1983.

Wilhelm, Paul. *Travels in North America, 1822–1824.* Translated by W. Robert Nitske. Edited by Savoie Lottinville. Norman: University of Oklahoma Press, 1973.

Wilkie, J. S. "The Idea of Evolution in the Writings of Buffon, Part II." *Annals of Science* 12 (1956):212–27.

Willey, Gordon R., and Jeremy A. Sabloff. *A History of American Archaeology.* San Francisco: W. H. Freeman, 1974.

Wilson, Alexander. *American Ornithology, or the Natural History of the Birds of the United States.* 9 vols. Philadelphia: Bradford and Inskeep, 1808–14.

———. "The Foresters, Description of a Pedestrian Tour to the Fall of Niagara in the Autumn of 1804." *Port Folio* 2 (July 1809):70–77.

———. "The Particulars of the Death of Captain Lewis." *Port Folio* 7 (1812):34–47.

———. "A Pilgrim." *Port Folio* 3 (June 1810):499–511.

Wilson, James. "Audubon's Ornithological Biography." *Blackwood's Magazine* 30 (July 1831):1–16.

———. "Wilson's *American Ornithology.*" *Blackwood's Edinburgh Magazine* 30 (August 1831):247–80.

Wilson, James Southall. *Alexander Wilson: Poet-Naturalist.* New York: Neale, 1906.

Wilson, Leonard. "The Emergence of Geology as Science in the United States." *Journal of World History* 10 (1967):416–37.

Wilson, William E. *The Angel and the Serpent.* Reprint. Bloomington: Indiana University Press, 1984.

Winterfield, Charles. [C. W. Weber.] *"American Ornithology." American Review* 1 (March 1845):262–74.

Wistar, Caspar. "A Description of the Bones Deposited by the President." *Proceedings of the American Philosophical Society* 4 (1799):526–31.

Woodward, David, ed. *Five Centuries of Map Printing.* Chicago: University of Chicago Press, 1975.

Wyeth, John B. *Oregon: Or a Short History of a Long Journey.* Cambridge, Mass.: Printed for John B. Wyeth, 1833; reproduced by University Microfilms, Ann Arbor, 1966.

Index

ABOUT THE AUTHOR

Charlotte M. Porter is an Assistant Curator, Florida State Museum, University of Florida, Gainesville, Florida. She received her A.B. from Bryn Mawr College and her M.A. and Ph.D. from Harvard University. This is her first book.